WET SCRUBBERS

WET SCRUBBERS
SECOND EDITION

Kenneth C. Schifftner, BSME
Technical Director, Compliance Systems International

Howard E. Hesketh, Ph.D., PE
Professor Emeritas, Southern Illinois University

TECHNOMIC
PUBLISHING CO., INC.

LANCASTER · BASEL

Wet Scrubbers

a TECHNOMIC*publication

Published in the Western Hemisphere by
Technomic Publishing Company, Inc.
851 New Holland Avenue, Box 3535
Lancaster, Pennsylvania 17604 U.S.A.

Distributed in the Rest of the World by
Technomic Publishing AG
Missionsstrasse 44
CH-4055 Basel, Switzerland

Printed in the United States of America
10 9 8 7 6 5 4 3 2 1

Main entry under title:
  Wet Scrubbers – Second Edition

A Technomic Publishing Company book
Bibliography: p.
Includes index p. 199

Library of Congress Catalog Card No. 95-62014
ISBN No. 1-56676-379-7

Welcome to the second edition of *Wet Scrubbers*.

It hardly seems that it was more than a decade ago that Dr. Howard Hesketh and I wrote the first edition of *Wet Scrubbers*. The purpose then was to provide an easy-to-read, easy-to-understand technical book on the design and application of gas cleaning technologies that use liquids. The original edition found its way onto the reference shelves of plant engineers, environmental engineers, regulatory personnel, and others who spend all or part of their time studying and/or solving air pollution problems. It has also been added to many libraries as a useful reference tool for students and others interested in pollution control. We have been very pleased to hear many favorable comments from those individuals who use *Wet Scrubbers* as a reference.

Our purpose in providing this second edition is to build upon that base and update a number of chapters to reflect the exciting changes that have occurred in the industry during the last decade. We have added to the theory of operation of wet scrubbers, included a section on new "hybrid" scrubber systems that use multiple control techniques, and added a section on irrigated fiberbed filters that utilize Brownian motion capture techniques for fine aerosol capture.

During the last decade, regulatory codes have become increasingly stringent, prompting designers to seek either better performance for their existing control equipment or to add new equipment of higher efficiency. Technology to control specific pollutants such as mercury compounds, hexavalent chrome, and dioxins have all been "developed" in the last decade in response to real or perceived needs to control these pollutants. Some of the processes "developed," such as flux force condensation scrubbing, were in actuality rediscovered. Coupled with cost constraints, the challenge facing personnel confronted with environmental compliance can appear daunting. A careful review of the technologies available, however, can lead to successful completion of such projects.

It can be said that air pollution control is a mixture of control technology, economics, health sciences, safety considerations, plant design, law, statistics, judgment, and, yes, beliefs. The air pollution control engineering environment, we feel, is regulatory-driven, based upon the popular belief that the public health should be protected at a reasonable cost. It is based upon the concept that, when we derive benefit from Nature, we should not destroy Nature in the process. Some say this effort should be tempered by applying "cost/benefit analysis" to these health issues. The response is that we already do. The marketplace performs a cost/benefit analysis when an individual or company selects a specific pollution control technology to use. It can be further argued, "What is the dollar value of a human life?" Is it not priceless? The air pollution control industry reacts by agreeing that human life is indeed priceless, but one should not have to go bankrupt to protect it.

Choosing the appropriate system and operating it properly goes a long way in this regard, and thousands of air pollution control systems in successful operation around the world are testament to the fact that one can protect the public health without excessive cost. In addition, pollution control engineers work daily to reduce the operating cost of control equipment. New products continually arrive on the scene offering efficiency improvements at reduced cost. Older, less efficient, more costly systems fade away.

This second edition of *Wet Scrubbers* will help personnel who are involved with air pollution control issues make important decisions regarding how pollution can be reduced in an economical manner and feel confident that they have made the proper decision. The contents of this book are intended to supplement other information the readers may have about the problem and to reach an educated conclusion about the best course of action.

PREFACE TO THE FIRST EDITION

The application of wet scrubbers to industrial sources is both an art and a science. The science is called "applications engineering." The art is called "experience."

In this monograph, we provide information of sufficient detail to permit plant engineers and those interested in air pollution control to select and properly apply commercially available technology, combining art with science. To make this task easier, we have endeavored to present basic background information, as well as specific examples, in a concise manner.

Each application section is summarized by an Applications Guide. These guides contain general information regarding each specific application, including expected pressure drops, types of equipment, possible trouble areas, items to avoid and design specifics.

We do not pretend to present the only "right way"; there are various means to the same end when it comes to air pollution control. Our examples are based on successful installations, with information contributed as much by those people and firms who use the equipment as by those who designed the systems. The use of equipment exposes its advantages and pitfalls beyond the picture painted by the salesman, his brochures or the company he represents. Use alone shows whether the design is proper for the application. Use reflects the true cost of an air pollution control system, not merely the initial cost. Proper use is the successful blending of application engineering science and the art of experience.

KENNETH C. SCHIFFTNER
HOWARD E. HESKETH

ACKNOWLEDGEMENTS

We extend our appreciation to the variety of contributors who provided photographs, literature, charts, graphs and commentary to substantiate our examples. We also wish to thank the Industrial Gas Cleaning Institute for its input regarding specific techniques examined in depth in this work.

We sincerely thank Patricia A. Schifftner and Robert P. Hesketh for their help in preparing materials and proofreading. In addition, the secretarial assistance provided by the Southern Illinois University at Carbondale is greatly appreciated.

Every working day, millions of cubic feet of industrial air are cleansed of hundreds of tons of particulate and containment gases using wet scrubbers. These specially designed scrubbing systems prevent contaminants from attaining levels which may be injurious to people, plant and animal life, and the environment.

The contaminants that these systems control would otherwise accumulate in our ecosystem, to be removed only at a far greater cost when dispersed than when concentrated at their source. A properly designed scrubbing system provides an important opportunity to control the contaminant effectively at a point of lowest cost and complexity.

The challenge for an engineer confronted with an air emissions problem involves a number of factors, the most important of which are:

(1) Knowledge of the available control techniques
(2) Knowledge of the emissions source
(3) The ability to apply the proper device to the source

Our efforts here have been directed to providing information that will aid an engineer in each of those prime areas. We have described many available control techniques, typical sources and means of applying those techniques to the sources. We have avoided detailed descriptions of process source characteristics, recognizing that even among similar industrial problems, wide variations will exist.

Our point of view is more that of air pollution control engineers than process engineers. By so narrowing our attention, we feel that much can be gained by sharing, in a concise manner, the information regarding practice rather than theory. We document that practice with illustrations, sample calculations and suggestions.

Few industrial marketplaces are as competitive as the sale of air pollution control equipment. Each manufacturer claims that his specific device is superior to "the other guy's." In the area of wet scrubbers, common physical forces acting in each vendor's design make a differentiation of products even more difficult. We have been fortunate to depict drawings and photographs of many known successful industrial installations using materials supplied by system vendors. At every turn, vendors responded openly to our request for information. We sincerely thank all contributors for their assistance.

Because it would take thousands of pages to document all industrial applications, we have instead carefully chosen a few instructive ones. By reading the explanations of the general theory of collection, gas absorption, system design and vessel configurations, the specific lessons learned may be applied successfully to many other applications as well.

Air pollution control is an important yet expensive obligation of the industrial producer. Our society, through its environmental laws, requires him to control potentially harmful emissions at their source. Unfortunately, he must bear the cost of control. Our work here is intended to help reduce that cost. Wish him well in his task, both for our sake and for the safety of future generations.

acfm = actual cubic feet per minute
dscfm = dry standard cubic feet per minute
FGD = flue gas desulfurization
gr = grains, 1/7000 of a pound
HETP = height equivalent of a theoretical plate
L/G = liquid-to-gas ratio, gal/1000 acf saturated gas
N_T = number of impingement trays
ORP = oxidation reduction potential
o.u. = odor units
ppm = parts per million (by volume if gaseous; by mass if liquid or solid)
scf = standard cubic feet
s/s = stainless steel
TPD = tons per day
Vp = velocity pressure   *column*
w.c. = "water closet," refers to height of a water column
w.g. = "water gauge," refers to height of a water column
ΔP = pressure drop
ϵ = efficiency, fraction
μm = micrometer or "micron"
ϱ_G = density of gas, lb/ft^3
ϱ_L = density of liquid, lb/ft^3

CHAPTER 1

Application Engineering Basics

1.1 INTRODUCTION

In the industrial world, scrubbers have been used for a long time. In 1836, for example, a packed tower absorber scrubber patent was issued. One hundred years later (in 1935) the English were removing 98% of the SO_2 in flue gas scrubbers. A particulate control scrubber was patented in 1901 extending the recognized use to both gas and particle control. Even incinerators, which can be good pollution control devices in themselves, may require scrubbers to enable the emissions to meet the regulations. Figure 1.1 shows the stack of a wet scrubber system on a city of Detroit municipal incinerator. Note the characteristic brilliant white scrubber plume that occurs during cold weather.

1.1.1 Scrubbing System Inputs

Application engineering for air pollution control equipment requires a knowledge of the process to which the system will be applied and of the specific equipment which will be utilized to control the emissions.

Regardless of the vendor's name, applications engineering begins with some basic calculations. These calculations serve to provide gas flow data from which liquid flow, pressure drop, droplet removal, evaporation rate, and equipment proportioning data are derived. The basic calculations involve two areas: (1) determination of carrier gas conditions, and (2) determination of pollutant characteristics.

In the first area, psychrometric data are used to determine the following information:

- saturated volume

1

FIGURE 1.1. Characteristic wet scrubber plume in cold weather.

- scrubber outlet temperature
- scrubber outlet gas density and pressure
- evaporation rate
- outlet enthalpy

In the second area, gas and/or particulate data are synthesized to provide the following:

- particulate size distribution
- pressure drop
- inlet concentrations
- required efficiency
- partial pressure of contaminant gases
- cooling effects (sensible)
- exothermic or endothermic reaction effects
- wear characteristics

The potential user of air pollution control equipment must provide to the designer the following information so that the foregoing may be calculated:

- inlet gas volume, temperature, humidity and pressure
- peak particulate and gas loading
- average particulate and gas loading
- inlet scrubbing liquid
- particulate size distribution
- means of disposal of contaminant

The preceding data may be obtained through the use of a stack test or by diagnostic testing. Diagnostic testing does not attempt to appear as a compliance test. Diagnostic testing is the extraction of data through system testing designed to produce the variables needed by the designer in the solution of the problem. It may involve particulate distribution analyses, photography of the stack or emissions source, material balances and the like.

The means of disposal of the contaminant must be investigated at the outset. If sewering of the contaminant is permitted, the local regulations must be investigated as to concentration limits, additive limits, or other constraints. If the waste must be treated, the plant process must be investigated to determine overlap processes that may be utilized in the cleanup system. This may involve the blending of a caustic flow from one part of the plant with an acidic scrubber wash water or the addition of the scrubber underflow at a specific point in the facility's treatment system. If no treatment system exists, the requirement must be considered by the designer. A system dedicated only to the scrubber may be imperative.

Many companies perform this data acquisition function themselves;

others hire testing or consulting firms to provide these data and a technology assessment (TA). The TA outlines the problem and investigates alternative technology currently in use in the solution of the problem.

From the input data, one can calculate the Carrier Gas Conditions. This procedure involves the determination of the specific humidity (weight of water vapor per pound of dry gas) and uses psychrometric charts to determine the saturation temperature. These charts also provide the saturated gas conditions, such as temperature, outlet humidity and outlet enthalpy.

Section 1.3 describes these basic calculations for reference in the solution of many industrial wet scrubber applications problems. Vendors are an alternate source of applications information; however, one must keep in mind that a vendor's purpose is a "sale." His solutions tend to revolve around his particular type of equipment. We must remember that there exist a number of workable solutions to your applications problem; however, the best one is the one you feel comfortable with, given your particular problem.

For those unfamiliar with the types of collectors available, we will present in the next few sections a summary of their characteristics, limitations and other details of design which may have a bearing on the type of design you may select for a given application.

1.2 SCRUBBER TYPES

Wet scrubbers may be divided into two major areas: particulate collectors and gaseous emissions control devices. The particulate collectors rely on inertial or electrostatic forces for collection of airborne particulates. The gaseous emissions control devices create large liquid-to-gas areas so that gaseous pollutants may be absorbed by the scrubbing liquid. Physical forces, such as pressure drop, are usually minimized in gas absorbers, whereas their effectiveness is maximized in particulate collectors.

We will discuss each category separately (though most scrubbers function concurrently as absorbers and particulate collectors, by design or by accident).

1.2.1 Particle Collection

Particulate collectors rely on inertial (Newtonian) forces for the collection of contaminants carried in a gas stream. Electrostatic forces are also utilized, but only to a minor extent in conventional collectors. Of the inertial forces used, the following predominate: (a) impaction, (b) interception and (c) diffusion.

In Section 1.3, we will discuss these forces as they relate to specific equipment designs. We would like to define our terms in this section so that a better understanding of the operation of existing equipment occurs.

1.2.1.1 IMPACTION

Impaction is the most prevalent means of particulate removal. The contaminant is accelerated and impacted onto a surface or into a liquid droplet. In this method the kinetic energy of the particle is used to penetrate the surface tension of the scrubbing liquid, allowing the latter to enclose the particulate, raising its density so that it may be inertially separated more easily. Some devices accelerate the particulate directly into a liquid film; others (spray scrubbers and venturis) produce a spray of water which permits collisions of the particulate onto the droplet's surface. As we will see in the following sections, the designer provides an environment conducive to these impaction collisions, thus increasing the probability of a collision and therefore the likelihood of particulate removal from the gas stream.

1.2.1.2 INTERCEPTION

Those particles which do not directly impact can be captured by interception. Here, the particles meet the droplets at angles far less than 90°, yet still have sufficient energy to cause the droplet to engulf them. Interception is droplet density–related. The chances for interception increase when the density of droplets increases. Devices such as spray-augmented scrubbers and high-energy venturis rely on creating high-density sprays of fine droplets in an effort to increase interception. Interception is most prevalent on submicron particulates which, given their low mass, tend to follow gas streams. Their kinetic energy levels are not great enough for impaction, but the characteristic small size and random motion encourage interception.

1.2.1.3 DIFFUSION

Diffusion is most noticeable in particles less than 0.5 microns (μ) where electrostatic forces begin to make an appearance. These particles migrate through the scrubbing liquid spray along lines of irregular gas density and turbulence. This diffusion action brings them into contact with liquid droplets in which there is little energy difference between the collecting liquid and the contaminant.

To encourage diffusion, the designer must allow proper residence times in the collecting area (venturi throat, impaction zone, impingement tray or the like) to permit this action to occur. Diffusion is most pronounced where there is a high temperature difference between the gas stream and the scrubbing liquid. The accelerated evaporation rate of the liquid encourages zones of varying density, causing the particle to pass from the more dense to less dense zone.

When intermolecular diffusion occurs, this is called diffusiophoresis. This action results from isolated and rapidly changing evaporation or condensation rates within the scrubbing liquid regime. These very weak forces are only applicable to extremely small particles, those less than 0.05 μ. On applications wherein a hot dry gas stream carries submicron particulates, diffusiophoretic forces can be used by preconditioning the gas stream using a spray quencher or similar device. This raises the humidity of the stream, encouraging the migration of the small particles to capturing droplets.

A related phenomenon is thermophoresis. In this area, the temperature difference (and hence energy level) across or between molecules in a gas stream tends to propel them to a lower (cooler) energy surface or droplet. Many designers of scrubbing systems spray cold makeup water into spray quenchers to enhance this effect. They recycle warm water only to those areas which have uniformly reached saturation. Thermophoretic and diffusiophoretic forces are more academic than practical when one *designs* a scrubber; however, they do exist and must be considered.

1.2.2 Gas Absorption

All scrubbers utilizing liquids for gas absorption rely on the creation of large liquid surface areas through the use of a variety of mechanical methods. These methods include hydraulic spray, impingement tray, bubble cap tray, sieve trays, packing (modular and dump-type), grids and a variety of combination devices all attempting to create this high liquid surface area in as small a volumetric space as possible. Techniques can be divided into two basic categories: (1) those which flow the liquid over some type of media (packing, meshes, grids, etc.) and (2) those which create a spray of droplets.

Units which incorporate a flow of liquid over a media offer to the gas stream the area or surface of the liquid not in contact with the media. Those which utilize spray rely on small droplet sizes in which the droplet's spherical surface is the exchange area. In the latter technique, the greatest surface area will occur when the greatest number of the smallest sized droplets occupy the given scrubber volume.

Units which rely on media are designed from data supplied by the media vendor. "Packing factors" have been empirically produced by these vendors to indicate the ability of the specific type of media to create a given exchange surface. Media which produce a high exchange area per unit volume usually involve small packing (1 in. or less) or are fibrous media.

Obviously, if one wishes to increase the surface area when only the area *not* in contact with the media is functional, one must reduce the net surface area of the media itself (make it smaller). Unfortunately, when the size of the media is reduced, the pressure loss through the packing typically increases because the openings between the media decrease. This creates a "diminishing returns" problem.

For an applications engineer, the challenge is to select a media combination which provides adequate surface area with minimum pressure drop. This is an iterative procedure in which a particular packing is selected and, using the vendor's published literature, the removal efficiencies and pressure drops are calculated, with subsequent comparison to other media types. Specific absorption calculations are beyond the scope of this work; however, the reader is encouraged to acquire the gas absorption literature of a number of media suppliers for reference.

The following *must* be considered for gas absorption:

(1) The smaller the media, the more the particulate present in the gas stream will be collected, perhaps leading to pluggage.

(2) The larger the media, the greater the need for proper liquid distribution. Smaller packing media tend to distribute the scrubbing media more readily. Large packing (2–3 in.) tends to permit channeling of scrubbing liquid rather than contribute to uniform distribution.

(3) Small packing is lighter (considering dump-type packing) and therefore is displaced by airflow. Small packing usually requires hold-down grids to prevent these shifts.

(4) Modular packing presents two surfaces for absorption but also two surfaces for corrosion and solids deposition. Fibrous packing can act as a filter, accumulating but not rejecting solids.

(5) Modular packing is heavier after use, given particulate buildup. Handling procedures should consider this weight increase.

(6) Absorbers should not be used where particulate exceeds 2–3 gr/scf (exception, spray towers or grid scrubbers).

(7) Tray-type towers must have perforations or minimum clearances adequate for the service. Particulate can easily plug small (3/16- or 9/32-in. perforations) openings.

(8) Moving disc trays (so-called valve trays) are not useful where the scrubbing liquid scales or where particulate tends to stick. They are best suited to water-soluble contaminant applications.

(9) Removable trays must be adequately secured to prevent "sneak-by" around the tray. The sections must be small enough to be safely handled.

(10) Tray scrubbers must have a uniform flow of liquid across the surface.

The losses are the sum of the dry pressure drop and the static head of water above each grid. Areas with less liquid will offer less resistance and therefore higher flow rates of air. These areas will plug and build up more readily.

(11) "End weirs" can be used to improve distribution of liquid in tray scrubbers. This acts as a dam behind which a uniform head of water may be maintained.

(12) Presprays on tray scrubbers must use full cone nozzles which overlap in coverage.

(13) Spray-type distributors on packed columns should have header pressures just sufficient to produce the desired spray pattern. Excessive pressure may displace packing.

(14) Injection-type packing support grids, in contrast to flat grids, should be used wherever possible. Improved distributions of both gas and liquid are obtained.

(15) Increasing liquid rates to an improperly designed packed tower will not improve efficiency unless the liquid is more dilute. Absorbers are surface area producers. The thickness of liquid flows does not contribute to increased areas.

Increasing the packing quantity or switching to a packing with a superior packing factor will improve performance.

(16) Increasing flow rates to an impingement scrubber will many times decrease removal efficiencies. The weight of water over each cap, baffle strip, or valve represents an energy potential which must be overcome.

Trays should be run at their design rates.

(17) Absorbers decrease in efficiency when greater quantities of contaminant gases are passed through; they typically do not reflect good turndown capabilities.

Multiple units are suggested. (Spray towers and some sieve trays are exceptions. Give maximum and minimum gas flow conditions to prospective vendors.)

(18) The liquid circuit of a gas absorbing system is at least as important as the absorber itself. Contaminant reaction products must be removed from the system, and reagents, if any, must be properly replaced.

(19) Absorption is not additive. It is more difficult to scrub, e.g., HCl and ammonia from a process stream than to remove each separately. Multiple stages are typically required—one for acid-forming species and another for base formers. Some systems also use oxidation stages wherein an oxidant is injected.

(20) Many absorbers benefit from warm scrubbing liquid. Always check

the solubility curves. Inverse solubility contaminants (like chlorine) benefit from cold scrubbing liquids.

(21) Consider the vapor pressure of the reagent used (if any). If you overdose, *it* could become the contaminant.

(22) On systems with settleable solid particulate, use an external decant recycle tank. Don't return it to the absorber.

(23) Free access to all working parts is a good investment in any absorber design.

Spray-type devices (spray towers or cyclonic scrubbers) must have uniform spray distribution and adequate droplet size. They are adversely affected by lowering of header pressures, which enlarges the droplets. The nozzles of these units are the points of greatest restriction and wear and therefore they must be easily accessible. Headers designed for external access, such as retractable ones, are superior to those with fixed internal headers, which demand operator entry into the vessel for service. Individual shutoff valves for nozzles or headers are suggested.

Spray-type devices experience reduced efficiencies if the solids content of the scrubbing liquid is permitted to increase beyond design levels. The solids reduce the effectiveness of atomization in most cases, and poor atomization results in larger droplets with reduced net surface areas per unit volumes. Solids are tolerable, however; they are even desirable in some applications such as SO_2 absorption using lime or limestone where they serve to keep the headers clean and to act as nucleation points for reaction product growth.

A spray tower commonly utilized a chevron or similar droplet control device. When the spray is produced, the smaller overspray can easily flow out of the unit unless it is controlled by a droplet eliminator. Sometimes this is so excessive that a sieve tray or similar coalescer is used to enlarge the spray for subsequent control by a chevron. The chevron presents an abrupt gas flow directional change which, through the inertial effects of the droplets, causes separation of gas and liquid. Face velocities of 850 ft/min for vertical flow and 1200 ft/min for horizontal flow are commonly used for sizing chevrons.

There are such a variety of absorbers that time and space do not permit a discussion of each. It can be noted, however, that the most efficient ones, regardless of type, are those which produce the greatest liquid surface area for a given scrubber volume, at the lowest pressure drop. One should avoid particulate matter by removing it prior to the absorber.

1.2.3 Scrubber Velocities

It is necessary to work within specific velocity limits regarding the vari-

TABLE 1.1. Typical Scrubbing System Velocities.

Inlet Velocities: 50–60 ft/sec
Throat Velocities (venturi scrubbers): 90–400 ft/sec
Tangential Velocities (cyclonic scrubbers of separators): 90–120 ft/sec
Chevron Vertical Flow Face Velocities: 550–850 ft/min
Chevron Horizontal Flow Face Velocities: 900–1200 ft/min
Stack Velocities: 30 ft/sec for saturation temperatures of 155–180°F; 40 ft/sec for saturation temperatures of 125–155°F
Spin Vane Separator Velocities: 1400–1600 ft/min (open area or area not occupied by blades)
Vertical Cyclonic Separator Velocities: 8–10 ft/sec for vessels up to 9 ft; 10–12 ft/sec for vessels 9–14 ft diameter
Vertical Spray Tower Velocities: 10 ft/sec
Entrainment Point: 12 ft/sec (above this point submicron spray will rise, not fall)
Packed Tower Velocities: 2–6 ft/sec for vertical; 4–8 ft/sec for horizontal
Mesh Pad Face Velocity: 850 ft/min
Drain Velocities (liquid): 1–3 ft/sec
Header Velocities: 6–10 ft/sec
Quencher Velocities: 50 ft/sec for hydraulically atomized spray; 55–60 ft/sec for air atomized spray

ous scrubber designs. Too low a velocity results in inadequate absorption and/or particle removal due to low contact/turbulence. Too high a velocity causes flooding, carry-over and other adverse effects. Many scrubber velocities are proprietary. Therefore, Table 1.1 lists typical values that can be used as guides in calculations, as discussed in detail in Section 1.3.9.

1.3 BASIC CALCULATIONS

1.3.1 Use of Gas Laws

Wet scrubbers are mechanical devices that often perform chemical functions. They remove particulates through the application of physical forces, and cleanse the air of gases through mechanically aided chemical reactions.

All scrubber applications, regardless of size, begin with some simple calculations. The devices are physically sized by the gas and liquid flow characteristics of the contaminant stream; therefore, accurate gas volumes are needed. They evaporate water in most instances; therefore, evaporation demands are required. They receive their motive force through the use of fans or other prime movers; therefore, outlet volumes are mandatory. They usually discharge to the atmosphere; therefore, stack gas conditions are needed. Calculations adhere to the precepts of the Ideal Gas Laws.

There are few mysteries in scrubber volume determinations. Let us take a typical inlet condition and calculate some basic parameters:

Gas inlet volume	10,000 actual cubic ft/min (acfm)
Temperature	300 °F
Water vapor	20% by volume
Gas outlet volume	unknown
Temperature	unknown
Evaporation	unknown

Calculate the gas conditions by determining the mass flow of the constituents in the gas stream—the dry gas and the water vapor. Use an approximate molecular weight of air at 29 and that of water at 18. For this purpose, calculate the gas conditions at 70°F wherein a lb-mol occupies ~~379~~ ft³. *385* Treat the mixture as an ideal gas, following the Ideal Gas Laws; therefore, calculate the inlet dry gas flow rate as:

$$(V) \frac{29}{\underset{385}{\cancel{379}}} \frac{(460 + T_{70})}{(460 + T_{in})} \left(1 - \frac{(\% \text{ water vapor})}{100}\right) = \text{lb/min} = m_g \text{ dry gas}$$

$$(V) \frac{18}{\underset{385}{\cancel{379}}} \frac{(460 + T_{70})}{(460 + T_{in})} \frac{\% \text{ water vapor}}{100} = \text{lb/min} = m_v \text{ water vapor}$$

where

$$V = \text{inlet volume (acfm)}$$
$$T_{70} = 70°F$$
$$T_{in} = \text{inlet temperature, °F}$$
$$\% \text{ water vapor} = \text{vol \% of water vapor in the inlet stream}$$

This results in:

$$(10,000) \frac{29}{\underset{385}{\cancel{379}}} \frac{(460 + 70)}{(460 + 300)} (1 - 0.20) = 426.8 \text{ lb/min dry gas} = m_g$$

$$(10,000) \frac{18}{\underset{385}{\cancel{379}}} \frac{(460 + 70)}{(460 + 300)} (0.20) = 66.24 \text{ lb/min water vapor} = m_v$$

1.3.2 Use of Psychrometric Charts

Psychrometric charts, such as those compiled by Dr. O. T. Zimmerman

[1], provide plotted data based upon gas temperature and humidity. This humidity, H, is an indication of the amount of saturation of the gas stream. In a continuation of the example, this number is calculated as:

$$H = \frac{m_v}{m_g} = \frac{\text{lb/min water vapor}}{\text{lb/min dry gas}}$$

In our example,

$$H = \frac{66.24}{426.8} = 0.155 \frac{\text{lb}}{\text{lb}}$$

In the psychrometric chart shown in Figure 1.2 [1] the ordinate is the humidity, H. The abscissa is the incoming gas temperature.

1.3.3 Adiabatic Saturation Temperature and Humidity

When a dry or partially dry gas accepts liquid vapor without a net transfer of heat, this process is called *adiabatic saturation*. Nearly all wet scrubbers rely on adiabatic saturation to cause a drop in the gas temperature and an increase in the flow water vapor content. Adiabatic processes exhibit no net heat transfer. (Further cooling through the use of a cold scrubbing liquid is called *sensible cooling*, since the scrubbing liquid accepts the sensible heat of the gas and water vapor stream until equilibrium is reached.)

Thus, as water vapor is added to the gas stream, the scrubber conditions follow the adiabatic saturation lines, gradually decreasing in temperature and increasing in water vapor content, while the total energy remains constant.

Continuing the example, look at the inlet gas temperature on the abscissa (300°F) and follow the vertical line upward until it intersects the line delineated by our H value (0.155). If we follow the adiabatic saturation line toward the 100% saturation curve, we find a saturation temperature for this particular mixture at 148°F.

This means that an introduction of a properly mixed stream of scrubbing liquid to this system will produce an outlet temperature of 148°F if saturation occurs. Partial saturation will result in a temperature above this point; indications of sensible cooling through the scrubbing liquid or from the system to its surroundings will result in a lower temperature.

But what volume do we have? How much water have we evaporated? Table 1.2 shows that at a temperature of 148°F each pound of incoming dry air produces a saturated mixture which occupies 20.49 ft^3. Thus, if we take

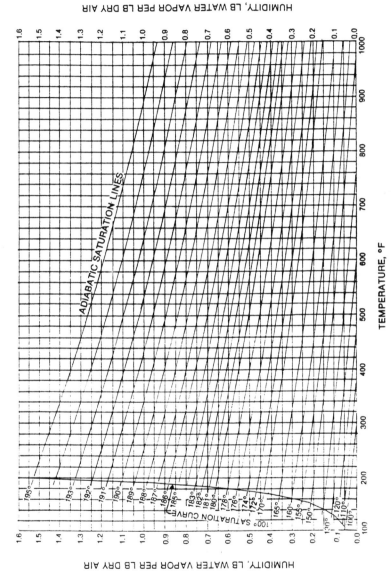

FIGURE 1.2. High temperature psychrometric chart [1]: adiabatic saturation lines and 100% saturation curve; temperature range, 100 to 1000 °F; pressure, 29.921 in. Hg; low humidity range.

13

TABLE 1.2. Volume of Saturated Air-Water Mixtures
near One Atmosphere Pressure.

Temperature, °F	Saturated Volumes, ft³/lb Dry Air
50	12.95
60	13.30
70	13.65
80	14.10
90	14.55
100	15.05
110	15.75
120	16.55
130	17.55
140	18.85
150	20.90

the inlet mass flow of dry gas (m_g) or 426.8 lb/min and multiply it by 20.49, we will obtain the outlet gas volume:

$$(m_g)(V_m) = \text{saturated volume (acfm)}$$
$$V_m = \text{specific volume of saturated mixture}$$
$$m_g = \text{dry gas flow}$$
$$(426.8)(20.49) = 8745 \text{ acfm}$$

Figure 1.2 gives saturation humidity as 0.20 lb of water vapor present for each lb dry air at 148°F.

Multiply inlet dry air mass flow rate by the saturation humidity to obtain the outlet water vapor flow rate:

$$(426.8)(0.20) = 85.36 \text{ lb/min water vapor}$$

At the inlet there were 66.24 lb/min water vapor. The difference is the evaporation quantity.

$$
\begin{aligned}
85.36 &= \text{outlet water vapor mass flow} \\
-66.24 &= \text{inlet water vapor mass flow} \\
\hline
19.12 &\quad \text{lb/min evaporated}
\end{aligned}
$$

Thus, to reach saturation we must effectively evaporate 19.12 lb/min of water. If we don't provide at least this flow, we will not saturate the gas, and our outlet temperature will be higher than 148°F.

These data show much more; consider Figure 1.2 again. For any given inlet moisture content, the higher the inlet humidity, the higher the saturation temperature. Look what raising the inlet humidity at a 300°F inlet temperature does. If we go from 20% water vapor by volume ($H = 0.155$) to 40% by volume ($H = 0.4138$) the saturation temperature jumps from 148°F to 172°F.

The applications engineer looks at the inlet gas conditions to show important trends. Some of these conditions, along with scrubber effects, are summarized below. These data are useful for ventilation concerns as explained in *Industrial Ventilation* [2].

(1) High inlet humidities yield high outlet saturation temperature, and, therefore, larger vessels for a given flow of inlet air.

(2) Processes which go from low inlet humidity to high inlet humidity often require variable (adjustable) scrubbers.

(3) Systems (such as dryers) that produce high saturation temperatures emit higher quantities of water vapor, and therefore are suspect for producing entrainment (droplet carry-over).

(4) High-inlet humidity gases produce dense plumes if emitted to the atmosphere.

(5) Low-inlet humidity gas saturates quickly, and therefore aids in scrubbing.

(6) Low-inlet humidity streams have typically higher evaporative losses; therefore, they require slightly higher liquid-to-gas ratios (L/G) since part of the flow is evaporated.

(7) Low-inlet humidity systems benefit from makeup (to replace evaporative losses) in the functioning part of the scrubber (throat of a venturi, cyclonic sprays of a cyclonic unit, etc.).

(8) High-inlet humidity scrubbers benefit from using cold scrubbing liquids (condensation scrubbing).

(9) High-temperature, low-humidity applications are, functionally at least, the easiest to design. High-humidity, low-temperature systems can be the most difficult.

(10) High-humidity systems sometimes benefit from dilution air to "temper" the inlet conditions.

(11) The "worst system" is one which fluctuates between the extremes of temperature and humidity, regardless of process application.

1.3.4 Pressure and Elevation Factors

Our calculations have been for air at atmospheric pressure. Most systems run in variance to atmospheric conditions; therefore, corrections must be made.

Gas density is reduced by physical elevation of the equipment or by inlet depression (suction). The psychrometric charts contain 100% saturation curves for various barometric pressures. The procedure we described is the same; however, the corresponding 100% saturation curve is used.

For inlet depression, the gas inlet volume is typically corrected first before further calculations. Inducing a suction on the inlet of a scrubber reduces the gas density, which conversely increases its apparent volume. A 1000 cubic ft/min (cfm) source would be increased in apparent volume once it reached the scrubber if an induced draft fan were used after the scrubber.

Since atmospheric pressure is approximately 407 in. of water column (w.c.), we simply expand our standard volume (V) by the factor:

$$K = \frac{407}{407 - \text{inlet suction}} \text{ (in. w.c.)}$$

$$\text{corrected volume} = V \times K$$

Conversely, inlet pressures reduce the apparent volume:

$$\frac{407}{407 + \text{inlet pressure}} \text{ (in. w.c.)}$$

Corrections for elevation are similar. One simply takes the barometric pressure at sea level and divides it by the pressure at the plant location. This factor times that volume of gas at sea level gives its "inflated volume" as the pressure of the atmosphere is reduced.

Knowing the volumes at given points in the scrubber, we can calculate superficial (free space, or open area) pressure drops. Manufacturers all have specific velocities for their units, some of which are proprietary. However, use the data in Table 1.1 as a guide and proceed with the calculations from there.

1.3.5 Particle Characterization and Removal

Particulate size distributions are extremely important in obtaining accurate energy requirements in scrubber design. Typically, the aerodynamic mean diameter (measured in microns) is obtained using a cascade impactor. Such a device, an Anderson system, is shown in Figure 1.3. These devices separate the dust into size fractions using a series of impaction stages. From

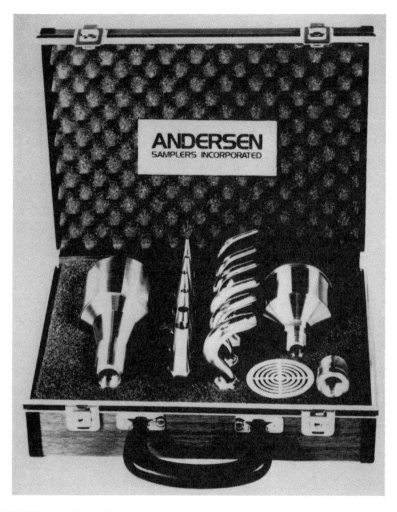

FIGURE 1.3a. Particulate size determination apparatus. Component parts (courtesy Andersen Samplers, Inc.).

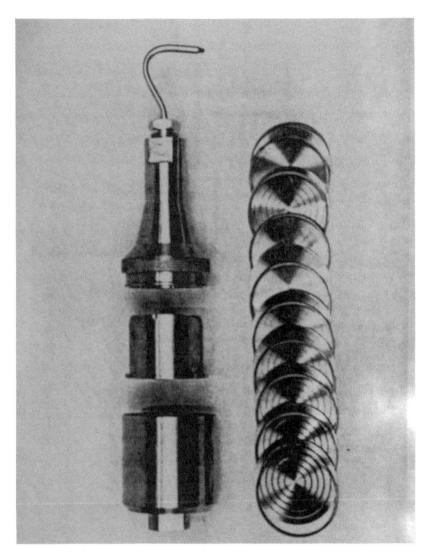

FIGURE 1.3b. Particulate size determination apparatus. Breakdown of the impactor (courtesy Andersen Samplers, Inc.).

these data, a particle size distribution by mass can be established. In extremely dusty samples, one should use a device such as the Anderson high-capacity impactor shown in Figure 1.4. The cascade impactors typically have 7 to 15 stages and are restricted to about 10 mg of dust sample per stage. The high-capacity system has only 3 stages plus a filter, but can sample up to several grams per stage. These devices can operate up to 1500°F. Some cascade impactors size dust that is as small as 0.03 μ.

Assume that from test data the following dust sample size distribution is obtained:

Aerodynamic Diameter, μ	Less than Stated Diameter, % by Mass
110	100
40	50
10	20
5	10
1	5

This tells the application engineer that 5% of the particulate is submicron (difficult to remove), whereas 95% is above 1 μ.

A full 50% of the particulate is 40–110 μ, quite easily removed by most low-energy wet particulate scrubbers. The range of 10–40 μ comprises 30% of the sample (50% less than 40 μ and 20% less than 10 μ). The range of 5–10 μ comprises 10% of the sample, 5% of the sample is 5–1 μ, and another 5% is less than 1 μ.

Going to efficiency curves for each particular product line, a determination can be made as to the efficiency required for a given outlet condition. Thus, if the inlet gases contain 10 grains (gr)/dscf (dry standard cubic feet) and the outlet has 0.07 gr/dscf, the efficiency would have to be:

$$E = \frac{10 - 0.07}{10}(100) = 99.3\%$$

Assume that a venturi scrubber, for which the data in Figure 1.5 were obtained, is to be used. Using the size data from above and the curve at 11.5 in.

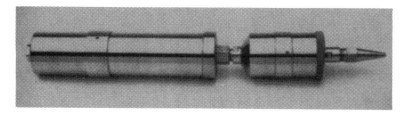

FIGURE 1.4. High-capacity impactor particulate size determination unit (courtesy Andersen Samplers, Inc.).

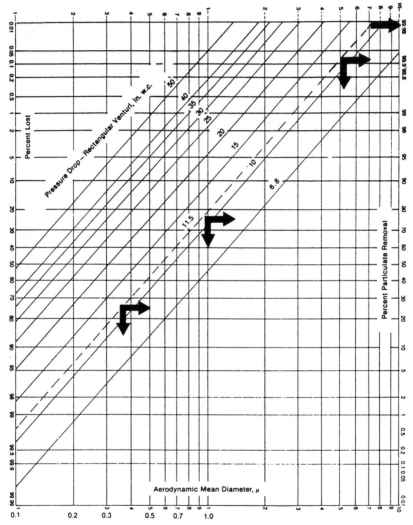

FIGURE 1.5. Pressure drop example for Section 1.3.5.

20

water column (w.c.) pressure drop (ΔP), the overall efficiency by mass of this scrubber at 11.5 in. w.c. ΔP on this dust would be (rounded to one significant figure):

Size, μ	% in Size Group	×	Efficiency	=	Net Efficiency, %
+10	80		0.99		79.2
5–10	10		0.99		9.9
1–5	5		$\dfrac{0.75 + 0.99}{2}$		4.4
<1	5		0.25 avg.		1.3
			Total		94.8

Since 99.3% efficiency is required, this device will be inadequate at 11.5 in. w.c. It is then necessary to choose a higher ΔP curve. Using a trial and error procedure it is possible to establish the necessary pressure drop. Variations of this procedure can be found, for example, in Hesketh [3].

1.3.6 Pressure Drop

As can be seen from the previous section, pressure drops (ΔP) in wet scrubbers can be related to control efficiency. Pluggage or erosion of a scrubber can affect ΔP, but it is assumed that these and other mechanical malfunctions are not permitted to exist. Figure 1.6 gives particle mass percent loss at a 50% level (i.e., cut diameter) for a rectangular-throat venturi as a function of pressure drop.

Once pressure drop is determined, it is necessary to calculate the volumes of gas in each area of the scrubber to properly size the system. Appropriate notation must be made for suction and pressure to account for correct gas volume changes.

A procedure for establishing venturi pressure drop is given by Hesketh [3]. In English units, pressure drop, ΔP, in inches w.c., is:

$$\Delta P = \frac{V_T^2 \varrho_G A_T^{0.133} L^{0.78}}{1270}$$

where

V_T = saturated gas throat velocity, ft/sec
ϱ_G = saturated gas density, lb/ft^3
A_T = throat area, ft^2
L = liquid-to-gas ratio, gal/1000 acf saturated gas

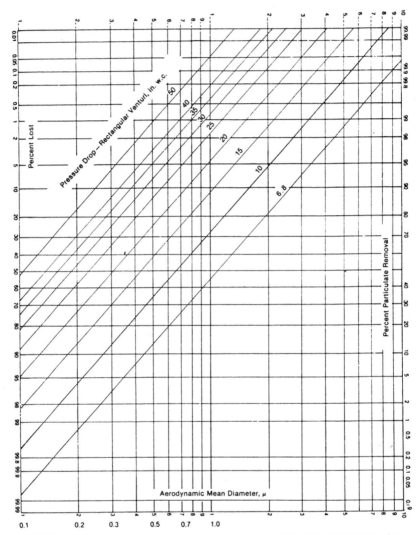

FIGURE 1.6. Venturi scrubber losses as a function of particle size and pressure drop.

22

Pressure drops for other scrubber types are presented in Section 1.4 with the appropriate scrubbers.

1.3.7 Solids Concentration

All scrubbers are affected by the solids concentration of the incoming gas stream. Spray-type scrubbers, which use high liquid velocities through small orifices, are particularly susceptible to plugging and wear at these points; therefore, we need to know the maximum particulate concentration in the scrubbing liquid. This is determined by acquiring the inlet dust grain loading (inlet particulate dust concentration measured in gr/scf). If the only datum is process emissions rate (lb/hr), congratulations, for this is the number we really need. If only gr/scfm is given, convert, using:

$$\frac{(\text{inlet grain loading in gr/scf})(\text{gas flow rate in scfm})}{7000 \text{ gr/lb}}$$

$$= \text{lb/min of dust}$$

Assuming our scrubber removes all of this dust, our recirculating liquid must contain this contaminant.

Take the design liquid recirculation rate (measured in gal/min) and divide by 8.3 lb/gal. This gives the *clean* water flow in lb/min. Now divide the contaminant lb/min by the clean flow lb/min, and you have the approximate dust concentration in the scrubbing liquid in weight % (wt %).

$$\frac{\text{lb/min dust}}{\text{lb/min water}}(100) = \% \text{ solids (wt)}$$

The concentration is usually limited to 20–30% for a venturi scrubber removing high–specific gravity dust to 0.5–1% for a spray scrubber with nozzle openings of 3/16 in. or less. The recirculation rate may have to be supplemented by bleeding the contaminated solids from the stream.

When we calculated the evaporation rate in Section 1.3.3, it was determined that 19.12 lb/min of water was required to saturate our example gas stream. If we must bleed away a fixed quantity of liquid to maintain proper solids concentration levels in the scrubber, we must make up that quantity plus the evaporation rate. Our calculation is:

$$\text{bleed} = \frac{\text{dust loading (lb/min)}(100)}{(\% \text{ solids})(8.3 \text{ lb/gal})} = \text{gpm}$$

$$\text{makeup} = \text{evaporation rate} + \text{bleed}$$

On hot-inlet (500°F or above) systems, the makeup is introduced into the inlet of the scrubber; on others, it goes into the sump; on units with after-coolers, it goes into the last cooler stage.

1.3.8 Quencher Calculations

We already calculated evaporative losses for a hypothetical system. A quencher utilizes evaporative cooling to reduce the outlet gas temperatures of high-temperature processes so that the scrubber following the quencher may reliably achieve saturation. If it doesn't, outlet dust loadings may be excessive and the operation of the scrubber may be adversely affected. This is especially true of short-contact venturi scrubbers.

Many quenchers are over-designed. In most cases, all that is required is to introduce the required evaporation rate as a spray into the quencher and to keep all wetted surfaces totally wet. Thus, many quenchers use a hydraulically atomized (or, for short quenchers, an air-atomized) nozzle to which 1 to 1.5 times the evaporative loss (in gpm) is sent.

In addition, 1–2 gal/min per foot of periphery of the quencher are introduced on a shelf or weir for covering the walls of the device. The quencher is mounted directly over the scrubber or has its own separate vessel.

Some quenchers use refractory tiles or baffles, on which a similar flow of liquid (at lower pressures, 5–10 psig or less) is applied. An example is metallurgical ore roaster gases or solid and liquid waste incinerator gases.

Spray-type quenchers generally follow these rules of thumb:

(1) The velocity should be 50 ft/sec or less.

(2) The contact time should be

	Inlet Temperature, °F	Contact Time, sec
(for hydraulically atomized sprays)	1500–2000	1.5–2
	1000–1500	1.0–2
	500–1000	0.75–1.5
(for air-atomized sprays)	1500–2000	1.0–2.0
	1000–1500	0.75–1.5
	500–1000	0.5–1.5

(3) Hydraulically atomized sprays should use 85–100 lb/in.² gauge (psig) water low in dissolved salts, if possible (to prevent spray drying).

(4) Air-atomized sprays should use clean water with 6–10 scfm/gpm of flow.

If the quencher discharges into a scrubber, the excess liquid will combine with the scrubbing liquid. One should take into account the dilution of the scrubbing liquid and the entry of hot makeup water. This is why quencher/venturi scrubbers are so often followed by aftercoolers (tray towers or packed towers). The outlet gases are not sensibly cooled in the scrubber; therefore, heat must in many cases be removed in the aftercooler.

We have depicted some of the basic calculations for wet scrubbing systems. Obviously there are a myriad of combinations which may occur in the application of wet scrubbers to industrial problems.

Now that the gas and liquid conditions are known, specific velocities may be calculated using the guidelines in Section 1.2.3 and in the following section.

1.3.9 Velocity Calculations

Scrubbing systems are mechanical devices performing chemical/mechanical functions. As mechanical devices, there exist definitive sizing parameters for each manufacturer's designs. These parameters are gas velocity-dependent and, therefore, structural in nature. Some specific component velocities are listed in Section 1.2.3. However, in general, the velocities required to size a scrubber may be subdivided into two areas — scrubber equipment velocities (such as gas entry and exit velocities), and removal velocities (venturi throat velocities, impaction and centrifugal velocities).

In the first group, typical velocities are described in Table 1.4. General ranges are given with higher velocities chosen for less abrasive applications (loadings of less than 1 gr/scf) as shown in Table 1.3 and Figure 1.7.

TABLE 1.3. Typical Scrubbing System Velocities.

Inlet Velocities:	45–80 ft/sec
Outlet Velocities:	With stack—30–35 ft/sec
	Without stack, to fan—50–60 ft/sec
Vane Eliminator:	1400 ft/min open area face velocity
Chevron:	500–600 ft/min (vertical flow)
	1100–1200 ft/min (horizontal flow)
Headers:	6–8 ft/sec
Drains:	1–3 ft/sec
Packed Towers:	2–6 ft/sec (vertical flow)
	6–10 ft/sec (horizontal flow)
Cyclonic Scrubbers:	90–120 ft/sec inlet velocity

*TABLE 1.4. Rectangular Throat Venturi C Factors
as a Function of Liquid Throughput.*

L/G, gal/1000 ft³ Saturated Gas	C
0 (air alone)	1050
4	1045
6	836
7	813
10	775
15	715
20	650
30	465
40	335
60	100

1.3.9.1 VENTURI SCRUBBERS

Removal velocities are more difficult to assess. This is especially true of
venturi scrubbers. Though various attempts have been made to derive an
equation for throat velocities, most manufacturers still use empirical data.

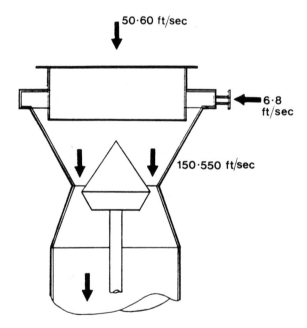

50·60 ft/sec

6·8 ft/sec

150·550 ft/sec

FIGURE 1.7a. Annular venturi velocities.

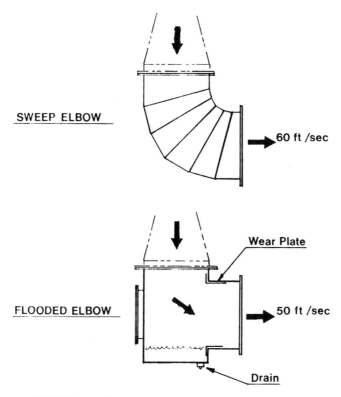

FIGURE 1.7b. Scrubbing system gas component velocities.

These data are derived from experience, trial and, unfortunately, error, and from pilot tests. In general, assumptions are made that:

- A mixture of scrubbing liquid and gas is in turbulent flow through the venturi.
- A given mixture of scrubbing liquid [liquid-to-gas ratio (L/G)] has reproducible hydraulic characteristics.
- The liquid is uniformly distributed across the area of the throat.
- The contaminant particulate is uniformly distributed across the throat.
- The throat has a fixed hydraulic loss at any given opening.

Some firms have computerized this information so that, when a given L/G, gas volume and pressure drop are entered, the throat velocity will be obtained. In real world situations, however, scrubbing liquid solids content, evaporation rates, nonuniform mixing and other factors come into play. We can make some useful generalizations, however.

The gas velocity in a venturi throat for a given pressure drop is related to the volume of gas passing through the throat, the density of the gas, the amount of liquid in contact with the gas, and a constant which corrects for the hydraulic losses of the particular throat or restriction. This can be presented as [a modification of Bernoulli's Equation (4)]:

$$A = \frac{Q}{C\sqrt{\Delta P/\varrho_g}}$$

where

A = throat area, ft²
Q = gas volume, acfm
C = correction factor, dimensionless
ΔP = pressure drop, in. w.c.
ϱ_g = gas density at saturation, lb/ft³

In this form, the venturi throat is considered as an orifice with certain inlet and outlet losses. It assumes a 30° approach and a 10–12° diverging section after the throat. In practice, however, approaches can be 50–60° and diverging sections 15–20° or, in some cases, nonexistent. The correction factor C includes:

- conversion factor for psi to inches of water (square root thereof)
- correction for turbulent flow (inlet and outlet) losses
- correction for L/G as it relates to hydraulic conditions of the throat

Table 1.4 gives C factors for various L/G ratios, for rectangular throats and 0.06 lb/ft³ gas density.

Decreasing the density of the saturated gas stream has a minor effect on the area of the throat. Increasing the density of the gas beyond 0.075 raises the C factor rapidly. At a density of 1.0 lb/ft³, the C factor is 2375 (at $L/G = 10$), given the increased resistance of the gas and liquid.

An empirical equation by Hesketh for predicting venturi scrubber pressure drop is given on page 21. Another equation developed by Hesketh for predicting venturi scrubber throat length is:

$$x_T = 328.582 v_T^{(0.02343L_G-0.8657)} \exp(-0.063L_G)$$

where

x_T = throat length, inches

v_T = saturated gas throat velocity, ft/sec
L_G = liquid-to-gas ratio, gal/1000 acf saturated gas

1.3.9.2 CYCLONIC SCRUBBERS

For cyclonic scrubbers operating at 4–6 in. w.c., Table 1.5 provides gas velocities (saturated gas volume divided by the cross-sectional area) for a variety of densities. (The L/G is 7.)

1.3.9.3 SPRAY AND OTHER SCRUBBERS

Spray-type cyclonic scrubbers are not noticeably affected by changes in L/G. Decreasing the L/G 20 to 7 for example, would decrease the velocities (the opening enlarges) by less than 10%. This is not true of high pressure drop venturis (40 in. and up). Here, a change in L/G from 20 to 7 for a 40-in.-w.c. throat would decrease the velocity from approximately 400 ft/sec to about 250 ft/sec. High-energy scrubbers are affected by the displacement influence of the water or other scrubbing liquids passing through the throat.

This phenomenon may be used to advantage in that the liquid rate can help control the throat pressure drop. It is also a potential problem in controlling a high-energy scrubber on a draft-sensitive system (such as a waste incinerator).

Annular venturis have center aerodynamic bodies which reduce the hydraulic losses to some extent. Typically, the throat area may be corrected by increasing the C factor by 40. It is suggested that specific vendors be contacted regarding their venturi throat systems.

Removal rates of particulate are determined by empirical data for each application. Removal efficiencies are logarithmic in nature: additional energy inputs in terms of velocity pressure present an ever-decreasing rate of removal. Going from 95 to 99% efficiency on certain applications can dou-

TABLE 1.5. Velocities for Cyclonic Scrubbers.[a]

Density	V at 4 in. w.c., ft/sec	V at 6 in. w.c., ft/sec
0.05	130	140
0.055	125	135
0.060	118	125
0.065	110	120
0.068	105	116

[a]Where scrubbing liquid passes through restriction with gas.

ble or triple the energy input. This is extremely important where submicron particulate predominates (metallic oxides, etc.). Many of the arguments regarding the Clean Air Act and its modifications found their basis in this increased energy input requirement. The reasons are complex.

1.3.10 Packed Towers

Vertical-flow packed towers operate at face velocities (area of packing support grid) of 2–6 ft/sec. The actual, saturated gas volume is used to determine the face velocity, though the gases may not be saturated until they are into the packed bed.

Horizontal-flow (cross-flow washers) towers are typically higher in throughput velocity. These range from 6–10 ft/sec depending on the liquid injection system and the inclination of the medium. Again, saturated volumes are used in the calculations of bed area.

1.3.11 Other

The previous sections of 1.3 provide some basic calculations procedures relative to scrubber design. Given the multitude of designs on the market, all cannot be covered here. Useful publications include the *Pollution Engineering Practice Handbook* [5] and others, such as those available from the Industrial Gas Cleaning Institute (IGCI), the U.S. EPA and publications listed in Section 1.7.

Fans and pumps are integral parts of scrubbing systems, but are not covered in detail in this monograph. For more information on fans, see *Fan Engineering* [4]. Pumps, which are the heart of the liquid flow system, are discussed in Walker's *Pump Selection* [6].

1.4 PARTICULATE SCRUBBERS AVAILABLE

Particulate collectors may be divied into the following groups, listed in increasing order of the ability to remove fine (submicron) particulate:

- spray scrubbers
- wet dynamic scrubbers
- cyclonic spray scrubbers
- impactor scrubbers
- venturi scrubbers
- augmented scrubbers

In each of these categories, *impaction* is a primary means of particulate removal for particles over 3 μ. In general, the better the device is at permitting repeated sources of impaction of contaminant onto a liquid droplet, the better the removal efficiency of particles above 3 μ. Gas scrubbers per se are not included in this section, as the list would be too long, but *all* wet scrubbers can absorb gas, as discussed in Section 1.2.2.

It has been demonstrated in actual operating scrubbers that particles above 3 μ tend to exhibit inertial effects (momentum, inertia, kinetic energy, etc.), whereas particles below this size tend to follow the gas stream, resist settling and resist inertial means of capture. These small particles are influenced by electrostatic forces, temperature gradients which draw them to cooler environments (thermophoresis), scrubbing liquid vapor pressure and liquid droplet size. In attempting to remove small particulate, the designer tries to create an environment with a high probability of one or more of the noninertial factors predominating. This is done by creating a high density of small liquid droplets in the regime of the contaminant, and maintaining this environment intact for a time sufficient to effect the required particulate removal. This environment is produced at the expense of energy in the form of spray nozzle hydraulic losses, velocity pressures or induced electrostatic charges.

Note that as particle removal increases, energy consumption increases as well (with a few exceptions to be noted later). Certain "novel" devices may prove to be more efficient for both particle removal and energy use, but they are not in common use now and are not included here.

1.4.1 Spray Scrubbers

These devices use hydraulically or pneumatically atomized streams of scrubbing liquid to remove particulate through direct impaction on large droplets, interception (accidental collisions) of smaller particulate on these droplets, or diffusion (where the particle is permitted to migrate to the scrubbing liquid droplet).

These devices are typically horizontal or vertical chambers in which banks of spray nozzles are oriented in a manner to cover the inner volume of the collector. These sprays must be mounted on the wall, or internally on headers. Given the high surface area produced by the liquid spray, the devices are good absorbers in many applications.

Typical efficiencies are 90% on particles greater than 5 μ and 60–80% on particulate 3–5 μ with a dropoff in efficiency to 40–50% or less on submicron particulate. Units which are used to remove difficult-to-wet materials (such as low-density plastic dusts and powders) are problem prone. Applications include the control of emissions from grinders, large pigment fugitive dust control, aggregate dryers (where nozzle plugging can be avoided), and knock-out applications in the fertilizer industry. Pressure drops are 2–4 in. w.c.

1.4.1.1 PARTICLE COLLECTION EFFICIENCIES

Hesketh developed equations for predicting particle collection efficiencies in spray towers [7]. The operating parameter ranges are for systems that use a "cool" or cooled inlet gas stream. Effective wet scrubbing consists of cooling (conditioning) hot gases before they enter the scrubber to minimize the adverse effects of negative diffusiophoresis and, if possible, to develop a positive flux force. If this is not carried out, the actual spray tower particle collection will be less than that predicted by the following equations.

1.4.1.1.1 Cocurrent Spray Tower Particle Collection Efficiency Model

Cocurrent spray tower particle collection efficiency is essentially only by impaction. Generalized particle collection in a two-stage cocurrent spray tower is:

$$Pt_{o,i} = \frac{(2.50)D_c^{0.574}V^{0.69}}{d_M^{1.60}[(L/G)L]^{0.393}}$$

where

$Pt_{o,i}$ = penetration due to impaction, percent
$\quad\ = 100 - E_{o,i}$
$E_{o,i}$ = impaction collection efficiency, percent
d_M = inlet dust mass mean diameter, μm
D_c = spray droplet mean diameter, μm
L/G = liquid-to-gas ratio, l/m³
$\quad L$ = effective collecting length below 2nd spray, ft
$\quad V$ = cocurrent gas velocity, ft/sec

This cocurrent spray tower equation is applicable for:

Variable	Range
d_M, μm	2.5–4.0
D_c, μm	700–900
L/G, l/m³	2.5–3.5
L, ft	35–45
V, ft/sec	18–22

The most significant variable is inlet dust particle size; second is gas velocity; third is collector droplet size; and the least significant are liquid-to-gas ratio and collecting length. Collection efficiency varies directly with particle size, L/G and L and varies indirectly with collector size and cocurrent gas velocity.

1.4.1.1.2 Countercurrent Spray Tower Particle Collection Efficiency Model

In a five-stage spray tower with countercurrent gas flow and following a quencher or prescrubber that releases only saturated particles, the particle collection is mainly by sweeping impaction of the falling droplets on the rising particles. A diffusiophoretic enhancement occurs due to condensation and growth of the smaller particles (i.e., those ≤ 1.5 μm). The resulting general equation for sweeping impaction including condensation growth is then:

$$Pt_o = \frac{(0.0568)D_c^{1.155}}{(L/G)^{0.606}L^{0.237}V^{0.126}}$$

where

Pt_o = overall penetration for the five spray levels, %
    = $100 - E_o$
E_o = overall efficiency, %
D_c = spray droplet mean diameter, μm
(L/G) = total scrubber liquid-to-gas ratio, l/m³
L = effective collecting length below lowest spray nozzles, ft
V = cocurrent gas velocity, ft/sec

This cocurrent spray tower equation is applicable for:

Variable	Range
D_c, μm	700–1420
L/G, l/m³	7.6–9.4
L, ft	8–16
V, ft/sec	9–11

The most important variable is spray droplet size; second is L/G; and third is L. Gas velocity change has little effect. Efficiency is directly related to L/G, L and V and indirectly to D_e. Note that in a countercurrent unit as L is increased, the ΔV decreases so the effect of increasing L to improve efficiency is not as significant as in the cocurrent spray tower.

1.4.2 Wet Dynamic and Cyclonic Scrubbers

Wet dynamic scrubbers and cyclonic spray scrubbers are considered together since their particulate removal efficiency is similar. Both of these devices remove particulate in the > 5-μ range at over 95% efficiency. They are shown in Figure 1.8. Submicron removal rates approach 60–75%.

Wet dynamic types include the sprayed fan designs, such as Ducon's UW-4® scrubber and American Air Filter's Rotoclone®. These devices feature a hydraulically atomized spray introduced into the inlet of a paddle wheel (or modified wheel) fan as shown in Figure 1.8b. Fan motor horsepower is expended in the movement of the air and in the creation of finely divided water droplets. Most of these devices precondition the air by spraying or passing the contaminant gas stream through a wetted knock-out area. This helps humidify the air stream, reducing the evaporation rate in the fan housing, and thus controlling the deposition of particulate on those surfaces. The fan discharge, now laden with droplets, is directed to a droplet eliminator which retains the liquid while permitting the cleaned gases to pass through.

Applications are predominantly for dust control in the mining industry, or for applications where the particulate sizes are above 5 μ. These devices are not useful on abrasive dust applications unless you can plan on purchasing new fan wheels frequently. They are inexpensive, and thus have a strong following.

Cyclonic spray scrubbers avoid the fan wheel spray by placing hydraulically atomized spray nozzle(s) into a cyclonic inlet separator. The gases enter tangentially into a vertical (though horizontal units do exist) cylindrical vessel wherein the spray is applied. The inlet duct to the cylinder is tapered such that an ever-increasing gas velocity is produced, along with an increasing droplet density. Impaction and interception are the predominant techniques employed. Figure 1.9 shows a cyclonic separator, which in itself could be used *after* any scrubber. With addition of horizontal and/or vertical spray headers, this separator could be a cyclonic scrubber.

The fan is typically mounted after the scrubber (clean air side) so the effects of abrasive wear are reduced. Since spray nozzles are used, strainers or decant tanks are frequently required in the liquid circuits, and must be added to the evaluated cost of these devices.

Typical applications include dust control in phosphate fertilizer plants, coarse dust control in grinding operations (sometimes followed by another

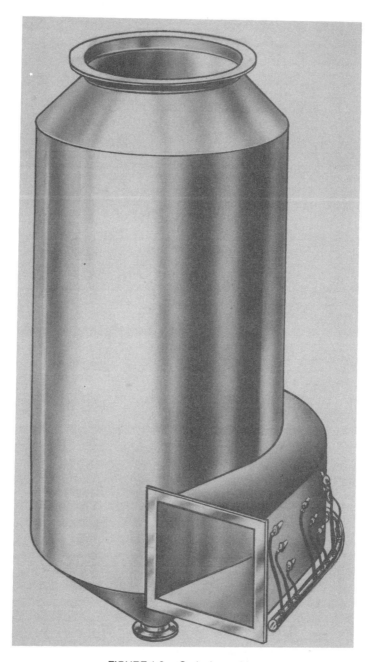

FIGURE 1.8a. Cyclonic scrubber.

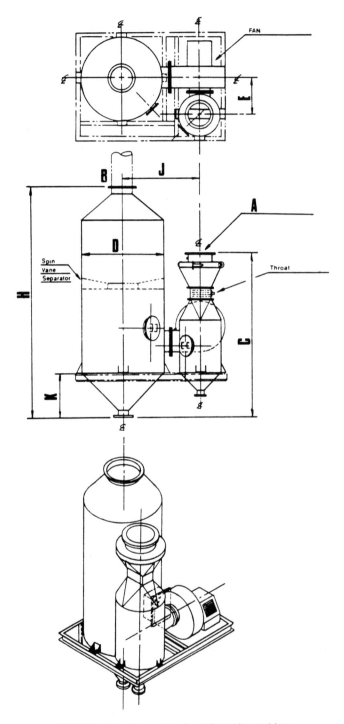

FIGURE 1.8b. Schematic of wet dynamic scrubber.

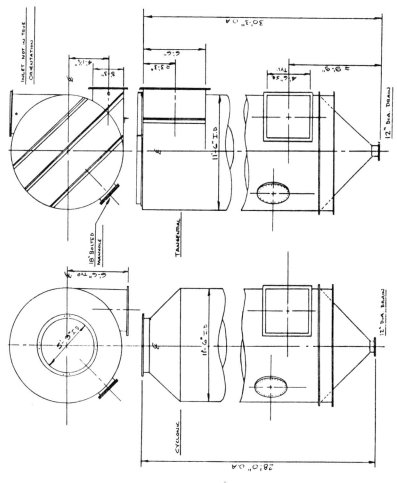

FIGURE 1.9. Schematics of cyclonic and tangential outlet separators.

37

scrubber), fugitive dust control, foundry shake-out particulate control systems, and other such projects where large visible particulate is emitted.

1.4.3 Impactor Scrubbers

Impactor scrubbers include impingement devices such as those manufactured by W. W. Sly and Peabody, and by Western Precipitation in the form of its Doyle-type unit. The Sly and Peabody designs incorporate perforated plates (of approximately 70% open area) whose openings are partially blocked by a target plate as shown in Figures 1.10 and 1.11. The gases are accelerated into the target plate and are washed from the surface by a constant flow of scrubbing liquid which passes over the plate. Western Precipitation's

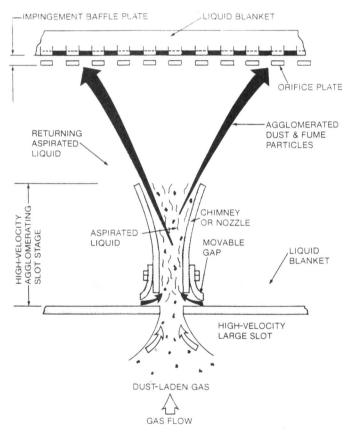

FIGURE 1.10. Adjustable slot plate impingement scrubber (courtesy Peabody Engineering Corporation).

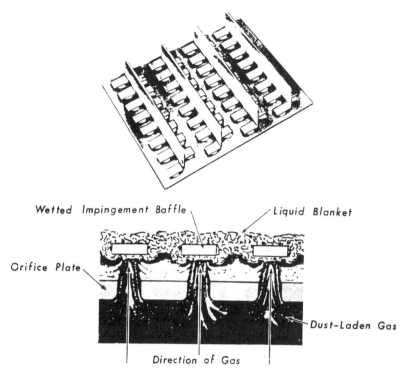

FIGURE 1.11. Impingement scrubber baffle plate (courtesy Peabody Engineering Corporation).

Doyle-type impactor accelerates the gases downward through a small annular gap directly onto a liquid surface. The effect in both cases is inertial impaction.

These devices have been used in a variety of applications since the 1940s. Impingement scrubbers were used on hundreds of lime kilns; some are still in service. Their excellent gas absorption properties lend them to use where halogenated gases must be scrubbed. They are also used on secondary metals applications and as cooling devices. They require dilute scrubbing liquids (usually less than 2% solids by weight) and are prone to plugging where the inlet grain loading exceeds 10 gr/scf.

The Doyle-type impactor (Figure 1.12) uses a prespray in many applications. The prespray conditions the entering gas stream, raising its humidity and typically lowering its velocity. These spray heads are oriented to completely cover the inlet area, but can be a source of problems if voids are created.

Impactor scrubbers have been successfully utilized on boiler applications where a large char is produced, as with bagasse boilers and bark boilers.

FIGURE 1.12. Doyle-type impactor on a boiler flue gas application (courtesy Western Precipitation Division of Joy Manufacturing).

The large size of the char is favorable to the inertial techniques used in this collector, yielding high efficiencies (over 98% removal) at pressure drops of only 3–8 in. w.c. Liquid ratios are low compared to venturi or spray devices since the scrubbing liquid is essentially static. Rates are determined by inlet grain loading and desired solids concentration in the sump. These devices are not effective on submicron particulate.

Impingement scrubbers can take on a variety of forms, each sharing the characteristic flat tray stages over which scrubbing liquid flows. These trays may be rigid, using baffle strips over perforated plates, or may include movable valve discs, or bubble caps, or even adjustable impingement trays. The technique involves producing a jet of contaminated air, its motion increasing the kinetic energy of the contaminated stream, which is directed at a target plate cleaned by a constant flow of scrubbing liquid. A froth of high surface area scrubbing liquid is produced in most cases, thus improving gas absorption properties. The target plate is the surface on which the contaminant impinges, hence the name.

This type of scrubber cannot tolerate high solids content scrubbing liquors, unless they are dissolved. Pressure drops are 1 in. w.c. dry drop and 1–2 in. w.c. wet drop per tray. These losses are additive; thus the typical unit has a pressure loss of 2–2 in. w.c. per tray. Adjustable trays can be operated at pressure drops of up to 20 in. w.c. Liquid rates are 2–10 gal per 1000 acfm, with higher rates used for gas cooling applications. In addition to the scrubbing liquid need, a liquid rate of about 2 gpm per ft² of active tray area (i.e., not including downcomers) is needed for the liquid seal. Trays are available in most malleable metals, polypropylene, PVC and other plastics.

Face velocities are as high as 14 ft/sec in some designs; however the static pressure loss becomes prohibitive. Optimum tray efficiencies are maintained when the F factor is about 1.8–2.0. The F factor is computed as:

$$F = V_G(\varrho_G)^{0.5}$$

where

V_G = superficial gas velocity, ft/sec
ϱ_G = saturated gas density, lb/ft³

Tray separation distance is usually about 24 in. Tray scrubber particle efficiencies are shown in Table 1.6.

TABLE 1.6. Percent Removal Efficiencies for Tray Scrubbers.

Particle Diameter, mm	Single Tray at ΔP = 3.5 in. w.c.	3 Trays in Series
0.6	~0	3
1.0	~0	48
1.5	31	74
2.0	56	87
3.0	86	97
4.0	98	100

1.4.4 Venturi Scrubbers

To remove greater quantities of submicron particulate (greater than 50% removal of particles less than 1 μ), one should consider using venturi scrubbers. These are gas-atomizing devices. Venturi scrubbers are mechanical devices which rely on shearing and impaction forces to break water into fine droplets. The goal is to produce an evenly distributed regime of fine droplets at high density. The high density raises the probability that a contaminant particle will impact on the droplet, enlarge through similar collisions, and become inertially separated (either through application of centrifugal force or through impaction on a medium such as a chevron). General venturi pressure drop–efficiency relationships are presented in Sections 1.3.5 and 1.3.6.

To make fine droplets from a large venturi throat is difficult. To make a uniform distribution of them is nearly impossible. Thus, one has an "overkill" situation wherein excess energy in the form of velocity pressure is added to the gas stream to correct these losses. Attempts have been made at spray augmentation of the scrubbers (introducing a spray nozzle liquid inlet just ahead of the throat), or by direct injection as in the American Air Filter Kinpactor® or Baumco-type venturi scrubber. In the latter, jets of scrubbing liquid are directly injected into the throat in fine streams, facilitating the breakup of the stream to fine droplets.

In practice, when a scrubber is operating at 45 in. w.c. pressure drop, one sees little improvement in removal rates on subsequent increases in pressure drop. Usually the venturi throat diverging section must be made longer, since the spray is produced not in the throat, but just beyond the throat (velocities of 450 ft/sec can prevail). With 1-ft-long throats, the contact time is less than 1/450 of a second in the throat, permitting minimal chances for particle contact in that zone.

On high-energy scrubbers, it is the *diverging* zone which should gain most attention. It is recommended that a diverging section of length at least four times the throat width be provided for scrubbers operating above 40 in. w.c. Many times, an existing scrubber can be improved by extending this diverging zone.

It seems that each manufacturer has several basic types of units: some are fixed throat and some are adjustable throat. The fixed throat category is comprised of round throat and rectangular throat scrubbers.

Round fixed throats are typically seen in high-pressure applications or where the materials of construction lend themselves to this shape, for example FRP (fiberglass-reinforced plastic) construction. If the throat is over 1 ft in width, a center spray or header is used to cover the center zone of the throat with scrubbing liquid; otherwise a void would be created at this area of maximum velocity.

Rectangular throats are usually limited to 18 in. in width for the same reasons. These throats are used where high abrasion is encountered, or where it is simply easier to build than a round unit. Such a unit is shown in Figure 1.13 with a cyclonic demister.

Adjustable venturis are more varied in design, yet each attempts the same

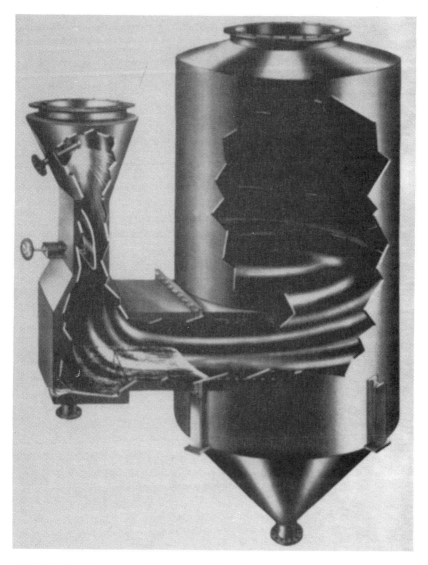

FIGURE 1.13. Peabody venturi scrubber with cyclonic demister (courtesy Peabody Engineering Corporation).

function—to permit operator adjustment of the throat area. The area of the throat determines the throat velocity. Curves are usually available from the manufacturer that show the throat velocity as a function of pressure drop and removal efficiency for a variety of particle sizes. Figure 1.14a is a dual-bladed venturi throat scrubber.

For adjustment, venturi throats may use damper blades, centrally

FIGURE 1.14a. Adjustable venturi throat. Note tangential liquid headers and damper blades (courtesy AirPol, Inc.).

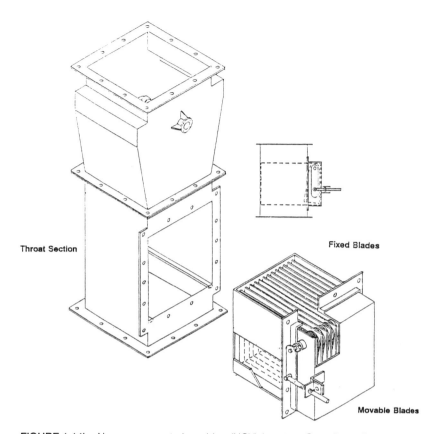

Throat Section

Fixed Blades

Movable Blades

FIGURE 1.14b. Narrow gap venturi scrubber (NGV) (courtesy Compliance Systems Int.).

mounted conical plugs (on annular venturis), wedges, discs or other devices. In general, devices which rotate on an axis mounted in the gas stream (damper blade types) adjust the area as to the sine of the angle of rotation, and thus exhibit sensitivity problems at higher pressure drops. A slight rotation of the shaft(s) will produce a drastic change in the pressure drop. This effect is not good when applied to a draft-sensitive device such as a boiler, dryer or fluidized device.

Annular types tend to behave linearly if properly designed, but are more difficult to maintain unless this problem is considered from the outset. They also have at least one seal on the push rod which actuates the plug. The center bodies or plugs can be mounted above the maximum constriction point of the throat or below. Any application wherein the venturi must "fail open" should have the plug *below* the constriction.

Unlike damper blade throats, which have throat widths in excess of 1 ft, annular types have widths of only a few inches, thus demanding reduced changes of direction of the gas stream. These units behave very well where throat velocities of over 200 ft/min are needed.

Another venturi scrubber design that seeks to limit the throat width to improve particulate capture is the Narrow Gap Venturi (NGV) scrubber (Figure 1.14b). This patented design uses a group of parallel fixed flat plates that divide the scrubber throat into zones. These zones are typically 1 in. to 2 in. wide. To make the unit adjustable, a pivoting set of similarly parallel blades are arranged such that they swing into the gap between the fixed blades. This arrangement is much like the design of an old tuning capacitor in a radio. As the movable blades swing into the throat zone, they decrease the throat area, increase the wetted surface, and decrease the throat gap further. This combination of events serves to change the hydraulic diameter of the throat and therefore change the throat pressure drop.

Either venturi type may have a wet or a dry "approach." This term applies to the hopper-like transition of the inlet to the throat area. Usually, the inlet spool extends into the venturi, thus avoiding "wet/dry line buildup." Wherever a surface is only partially wetted, the airborne dust will build up. Dry dust tends to seek areas of higher humidity, especially if the dust is hygroscopic. The inlet spool extension reduces this tendency.

By totally wetting the approach in dentist-bowl fashion or through the use of weirs, the wet/dry line can be controlled. Designs which rely on spray nozzles to accomplish this needed feature are inferior to designs of simple execution.

Venturi liquid rates range from 4 to over 100 gal/1000 acf. Spray-augmented venturis use lower rates, in general, than hydraulically atomized designs, but are more prone to nozzle plugging. Tangential inlet liquid velocities in annular designs are approximately 6–8 ft/sec for coverage of the dentist-bowl approach and header pressures of 5–10 psig are used for spray wetting. A variation of this type is the Peabody radial scrubber, shown in Figure 1.15.

Venturi scrubbers are in operation at over 100 in. w.c. pressure drops, at pressures of − 100 in. w.c. to hundreds of psig.

Variations include eductor designs such as those manufactured by Shutte and Koerting and Croll-Reynolds, and charge-enhanced designs such as the Pilat® scrubber.

1.4.5 Augmented Scrubbers (Electrostatically Enhanced)

In a high-energy venturi, we have reached the practical limits of inertial-based removal techniques because our target particles, the submicron species, behave as if they have little or no mass. Other forces have been shown to be more effective on these particulates; the most prominent is electrostatic attraction.

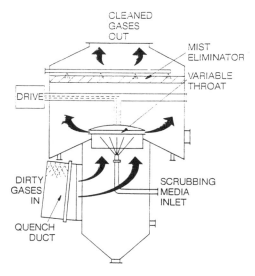

FIGURE 1.15. Schematic of Peabody Lurgi design up-flow radial scrubber (courtesy Peabody Engineering Corporation).

In this method, a conventional scrubber has its removal efficiency enhanced through the application of a positive charge to the incoming contaminant stream. Since the water tends to be electronegative, the small particulate is attracted to the water by electrostatic forces. Given the fact that water is a polar molecule, the tendency for water to act as a ground is not a pronounced one. This has led to developmental difficulties in the charge-augmented designs: one wants to charge the particulate but not the water.

Other designs, such as TRW's wet electrostatic precipitator, attempt to charge the water, introducing it as a charged spray wherein it migrates to a collecting electrode. Another design passes the contaminant stream through a carbon arc and subsequently to a packed column containing metallic packing media. In all of these designs, a charge is introduced to the particle so that it may be collected using electrostatic forces rather than inertial forces, which have little or no influence on the submicron species.

1.5 ABSORBER CALCULATIONS

1.5.1 Introduction

This brief discussion relative to absorber scrubbers is presented for the sake of completeness. The major portion will concern an example flue gas desulfurization system SO_2 absorber. Absorption is a mass transfer process and consists of transferring the pollutant gas from the gas stream into the liquid stream. The rate of mass transfer is dependent on liquid-gas interface area, the differences in concentration in the two phases (driving force) and the chemical species present (resolved as a mass transfer coefficient). Most of these data are provided by tables and figures.

1.5.2 Pressure Drop

It has been noted that surface area is directly related to the absorption of a gas. Table 1.7 lists various absorber tower packings by surface area and corresponding pressure drop per foot of packing. Pressure drop in both packed and spray towers would depend on the amount of absorbing liquid passing down the tower, but these values consider normal rates at about 80% of flooding.

1.5.3 FGD Absorption Example

Figure 1.16 shows a schematic SO_2 absorption system. It lists process conditions and physical properties for 90% SO_2 removal using sodium hydroxide (NaOH). The boiler flue gas has been quenched to 120°F as we receive it into the absorber, and the absorber packing is Munters type 12060.

Using the procedures of Section 1.3.1, the 500,000-acfm gas volumetric flow rate can be converted to a mass rate of 1.95×10^6 lb/hr (air at a molecular weight of 28.9 lb/mol is assumed). The liquid flow rate is

$$\left(\frac{15 \text{ gal}}{1000 \text{ acf}}\right)(500,000 \text{ acfm})\left(\frac{60 \text{ min}}{\text{hr}}\right)(8.34 \text{ lb/gal}) = 3.75 \times 10^6 \text{ lb/hr}$$

The flooding mass velocity is determined using Figure 1.17 by

$$x = \frac{L_M}{G_M}\sqrt{\frac{\varrho_G}{\varrho_L - \varrho_G}}$$

$$= \frac{3.75 \times 10^6}{1.95 \times 10^6}\sqrt{\frac{0.065 \text{ lb/ft}^3}{62.4 - 0.065}} = 0.062$$

Then from the figure, $y = 0.102$. Flooding mass velocity, G_F, then is

$$G_F = \sqrt{\frac{32.2 y \varrho_G \varrho_L}{F \mu_L^{0.2}}}$$

where F is the packing factor from Table 1.8.

TABLE 1.7. Comparison of Pressure Drop and Contact Area for
Countercurrent Absorber Packing [Gas—2250 lb/(hr ft²);
500 fpm Velocity; Liquid—2000 lb/(hr ft²); 4 gpm/ft²].

Packing Material	Available Surface Area, ft²/ft³	Pressure Drop in. w.c. per ft of Packing
Munters 12060	68	0.13
1-in. Koch Flexirings	65	0.90
1-in. Glitsch Ballast Saddles	65	0.80
1-in. Glitsch Ballast Rings	65	1.30
1-in. Intalox Saddles	63	0.75
1-in. Norton Pall Rings	63	0.90
1-in. Ceilcoat Tellerette	55	0.65
2-in. Maspak FN-200	43	0.75
1-1/2-in. Rashig Rings	40	1.60
1-1/2-in. Glitsch Ballast Rings	40	0.84
1-1/2-in. Koch Flexirings	40	0.75
1-1/2-in. Norton Pall Ring	39	0.75
2-1/2-in. Protak P-251	39	1.00
2-in. Ceilcoat Tellerette	38	0.30
2-in. Koch Flexirings	35	0.45
2-in. Protak P-252	34	0.82
2-in. Croll Reynolds Spiral-Pak	34	0.24
2-in. Glitsch Ballast Saddles	34	0.55
2-in. Intalox Saddles	33	0.50
2-in. Glitsch Ballast Rings	32	0.55
2-in. Norton Pall Rings	31	0.45
2-in. Heilex 200	30	0.45
3-in. Tellerettes	30	0.24
2-in. Rashig Rings	30	1.40
3-1/2-in. Koch Flexirings	28	0.22
3-in. Glitsch Ballast Saddles	28	0.32
3-in. Intalox Saddles	27	0.30
3-1/2-in. Norton Pall Rings	26	0.22
3-1/2-in. Glitsch Ballast Rings	26	0.22
3-3/4-in. Maspak FN-90	25	0.36
3-in. Heilex 300	23	0.27

TABLE 1.8. Packing Characteristics.

Packing Material	Packing Factor, F, ft²/ft³	Mass Transfer Diameter, D_M, in.	Surface Area, a, ft²/ft³
Munters 6560	58.8	0.224	123
Munters 12060	27.4	0.400	68
1-in. Rashig Rings	155	1.00	40
1-in. Intalox Saddles	98	1.00	63

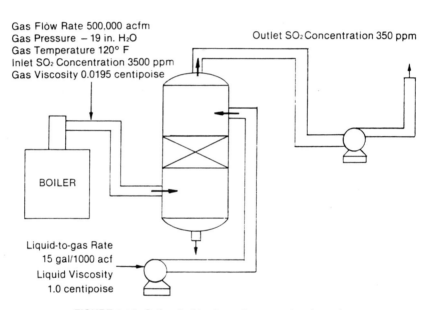

Gas Flow Rate 500,000 acfm
Gas Pressure − 19 in. H₂O
Gas Temperature 120° F
Inlet SO₂ Concentration 3500 ppm
Gas Viscosity 0.0195 centipoise

Outlet SO₂ Concentration 350 ppm

BOILER

Liquid-to-gas Rate
15 gal/1000 acf
Liquid Viscosity
1.0 centipoise

FIGURE 1.16. Sulfur dioxide absorption example schematic.

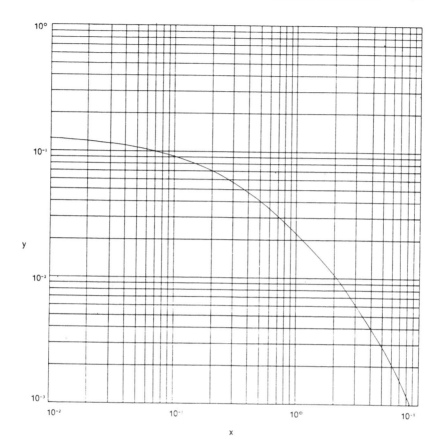

FIGURE 1.17. Flooding velocity graph.

Thus,

$$G_F = \sqrt{\frac{(32.2)(0.102)(0.065)(62.4)}{27.4(1)^{0.2}}} = 0.70 \text{ lb/(sec ft}^2)$$

Actual tower diameter is estimated at 80% of flooding velocity:

$$\text{area} = \frac{1.95 \times 10^6 \text{ lb/hr}}{(3600 \text{ sec/hr})(0.8)(0.70 \text{ lb/sec ft}^2)} = 967 \text{ ft}^2$$

$$\text{diameter} = \sqrt{\frac{4 \times \text{area}}{\pi}} \cong 35 \text{ ft}$$

Pressure drop, from Table 1.7, is 0.13 in. water per ft of packing. Height of packing is found using Figure 1.18 and mass transfer diameter, D_M, from Table 1.8:

$$G_M' = \frac{1.95 \times 10^6 \text{ lb/hr}}{967 \text{ ft}^2} = 2017 \text{ lb/(hr ft}^2)$$

$$L_M' = \frac{3.75 \times 10^6 \text{ lb/hr}}{967 \text{ ft}^2} = 3878 \text{ lb/(hr ft}^2)$$

$$Re_G = \frac{D_M G_M'}{\mu_G} = \frac{\left(\frac{0.4}{12}\right)(2017)}{(0.0195)(2.42)} = 1425$$

$$Re_L = \frac{D_M G_M'}{\mu_L} = \frac{\left(\frac{0.4}{12}\right)(3878)}{(1)(2.42)} = 53.4$$

$$Re_G^{0.67} \times Re_L^{0.23} = 324$$

Thus, from Figure 1.18, $J = 17.5$, where J is the mass transfer coefficient. Solve for the *overall* mass transfer coefficient, $k_G a$, using:

$$\text{gas constant} = R = 10.73 \text{ psi ft}^3/(\text{lb mol}°R)$$

$$\text{absolute temperature} = T = 120 + 460 = 580°R$$

$$\text{packing surface area} = a = 68 \text{ ft}^2/\text{ft}^3 \text{ from Table 1.8}$$

Schmidt No. $= Sc = \mu_G/\varrho_G D_{12}$, where D_{12} is SO_2 diffusivity of 0.604 ft^2/hr

$$= \frac{(0.0195)}{(0.065)} \frac{(2.42)}{(0.604)} = 1.20$$

Therefore

$$k_G a = \frac{J D_{12} a}{RT D_M} Sc^{-2/3} = \frac{(17.5)(0.604)(68)}{(10.73)(580)(0.4/12)}(1.20)^{-2/3} = 3.07$$

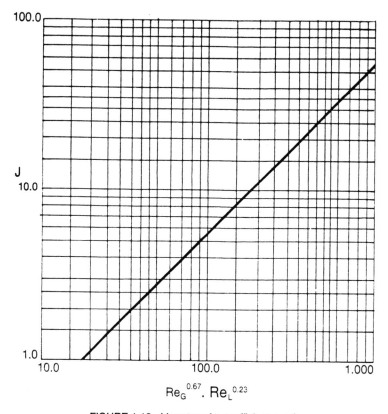

FIGURE 1.18. Mass transfer coefficient graph.

(Note that $k_G a$ will change with gas rate, type packing and other factors.) Packing height, H, then becomes

$$H = \frac{-G'_M}{k_G a M_G P_A} \ln \frac{y_o}{y_i}$$

where

M_G = molecular wt. of gas = 28.9
P_A = absolute pressure, psia = 14.0
y_o = SO_2 conc. out, ppm = 350
y_i = SO_2 conc. in, ppm = 3500

$$H = \frac{-2017}{(3.07)(28.9)(14.0)} \ln \frac{350}{3500}$$

$$= 3.74 \text{ ft}$$

Total pressure drop is

$$(3.74)(0.13) = 0.5 \text{ in. water}$$

At this point a word of caution should be made that applies to all absorbers. If the gas and/or liquid rates drop below the minimum or exceed the maximum design conditions, the column will fail to operate according to these described theories. For example, Figure 1.19 gives extrapolated theoretical and observed pressure drops for a sieve tray absorber at low gas flow rates. Note that agreement is good from the gas rate of 12.74 lb/sec (design) to 5.7 lb/sec (45% of design, and allowable). Below about 4.7 lb/sec, the pressure drop indicates that the liquid holdup has been lost and the column operation is no longer stable.

1.6 MIST ELIMINATION

1.6.1 Introduction

Mist elimination is an integral part of every wet scrubbing system. A failure at this point may often negate the entire scrubbing process and give rise to the statement that "scrubbers emit more particulates than are in the entering gas stream." It *can* happen, but obviously should not in a properly designed, operated and maintained system.

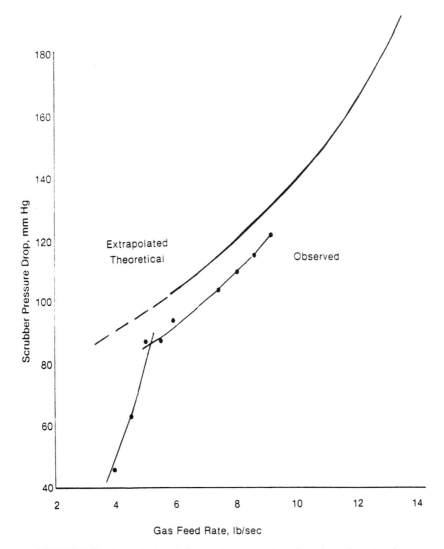

FIGURE 1.19. Theoretical and observed pressure drops in a sieve plate absorber.

Mist eliminators are also called entrainment separators and demisters. They serve to remove the liquid droplets from the exit gas stream whether the droplets are present because of entrainment, reentrainment, condensation or any other mechanism. Effective mist eliminators will remove 99 to 99.9% of the inlet liquid droplets.

1.6.2 Mist Eliminators

Mist eliminators operate mainly by inertial impaction and centrifugal force. However, interception and diffusion may also add to the droplet removal in the inertial impaction devices. The two most common types of inertial impactor mist elimination are the baffles and mesh, and there are numerous configurations of each. The two principal configurations are horizontal gas flow and vertical gas flow. The horizontal gas flow designs, which have recently been introduced into the U.S. from Germany and Japan, permit higher inlet gas velocities and liquid loading. The reason is that the separated liquid droplets do not fall back down into the rising gas if the gas flows horizontally through the eliminators.

Vertical gas flow eliminators usually are of the multipass chevron design such as those shown in Figure 1.20. Two to six passes are common, but three passes are usually sufficient. Of course, these zigzag baffle systems can also be used for horizontal gas flow. Note in Figure 1.20 that the number of passes is counted as the number of baffles in the unit. There is another method of counting passes as the actual turns of the gas stream within the device. In this procedure, the number of passes would be one less than that by the former method. Both systems are encountered in practice.

A modification of the continuous zigzag baffles is the slanted baffle demister shown in Figure 1.21. This improved system for vertical flow gas is inclined 30° from the horizontal and can operate at higher inlet velocities and liquid loadings. Another variation in vertical gas eliminators is the Munters Corporation Euroform® series eliminator shown in Figure 1.22. The walls contain chevron-shaped airfoils which serve to direct the liquid toward the ends of sections. The collected liquid droplets streaming from the edges of each section have less tendency to be reentrained by the rising

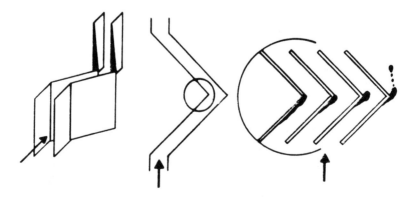

FIGURE 1.20. Chevron mist eliminator designs.

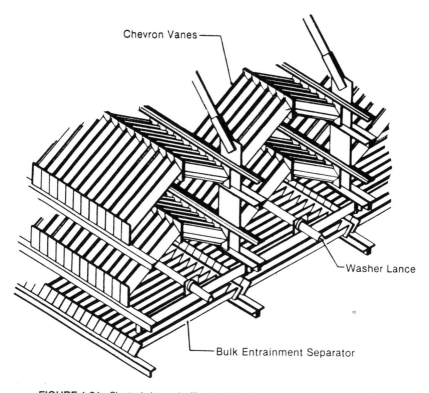

Chevron Vanes

Washer Lance

Bulk Entrainment Separator

FIGURE 1.21. Slanted zigzag baffle demister (courtesy Munters Corporation).

gas. This T271 eliminator is a high-efficiency, low liquid loading unit which often requires one of the nonclogging chevron mist eliminators, such as that in Figure 1.20, to be positioned below it.

Horizontal gas flow eliminators often include liquid phase separation chambers as shown in Figure 1.23. This reduces the carry-through of collected water.

Mesh-type eliminators are used for vertical gas flow when the gases contain no sticky materials that would plug the mesh. Figure 1.24 shows a typical mesh eliminator. Mesh thickness ranges from 4 to 12 in., with 6 in. being typical. The mesh pads may be installed at from 0 to 45° from the horizontal. Cylindrical mesh fiber packs and packed beds of fibers are also available, but they are less common for industrial scrubbers.

Cyclonic separators are good first-stage mist elimination devices. Coupled with gravitational force, these can remove large droplets. They require more space than the inertial impactors and they are less efficient in industrial sizes, but are less prone to pluggage. Figure 1.9 is a sketch of a cyclonic separator.

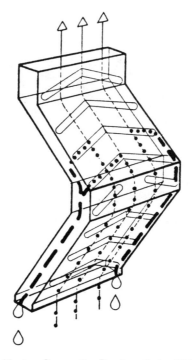

FIGURE 1.22. Munters Corporation Euroform Series T271 mist eliminator.

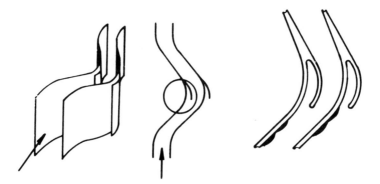

FIGURE 1.23. Phase separation chamber mist eliminator.

(a)

(b)

FIGURE 1.24. (a) Typical mesh pad mist eliminator with supports; (b) stainless steel wire mesh section (6 in. thickness).

1.6.3 Design Notes

Impaction-type mist eliminators should have wash provisions to keep them clean and to remove buildup. The wash usually is in the form of a fresh water spray directed from the underside or gas entrance side. The spray nozzles should operate at no greater than 25 psig to reduce the amount of fine droplets that could escape through the eliminator. Typical wash rate is 3 gpm per ft^2 of surface area.

Inertial impaction baffle or mesh eliminators should be securely tied or fixed in place so they do not move during startup or operation. Material of construction can be quite varied, although lightweight plastic and fiberglass (FRP) usually are advantageous. (FRP is good to 400°F.) Stainless steels and other suitable metals are also commmonly used.

1.6.4 Velocities

Velocities in general are capable of ranging from a maximum of 110% of design to a minimum of 60% of design. Efficiencies will decrease as velocities change in either direction. Reported velocities for well designed eliminators are given in Table 1.9.

Chevron mist eliminators with vertical gas flow have 99% and greater effectiveness for F_1 factors of 1.2 to 3.6. The F_1 factor is

$$F_1 = V_G(\varrho_G)^{0.5}$$

where

V_G = superficial gas velocity, ft/sec
ϱ_G = gas density, lb/ft^3

For horizontal gas flow, the F_1 ranges from 1.2 to 4.8.

TABLE 1.9. *Reported Mist Eliminator Velocities.*

Eliminator Type	Gas Flow	Gas Velocity, ft/sec
Zigzag	Horizontal	15–20 (inlet)
Zigzag	Vertical	12–15
Zigzag (30° from horiz.)	Horizontal	16–22
Cyclonic	All	100–130 (inlet)
Mesh	Horizontal	15–23
Mesh	Vertical	10–15
Tube Bank	Horizontal	18–23
Tube Bank	Vertical	12–16
Euroform T271	Vertical	10–20

Maximum velocity in mesh eliminators with vertical gas flow can be estimated in ft/sec by a Stokes' relationship and the Souders-Brown equation:

$$V_{max} = K \frac{\varrho_L - \varrho_G}{\varrho_G}$$

where

ϱ_L = liquid density, lb/ft³
K = 0.35 for mesh density of 9–12 lb/ft³
= 0.40 for <9 lb/ft³
= 0.30 for plastic and Teflon®* mesh

If the mesh is installed at some angle, θ, to the horizontal, replace K in the Souders-Brown equation with K_a, where

$$K_a = K + 0.3 \sin \theta$$

Table 1.10 is a comparison chart showing densities of various wire mesh eliminators.

1.6.5 Pressure Drop

Pressure drop in mist eliminators is low, ranging from about 0.5 to 1.0 in. w.c. under normal conditions and 0.2 to 1.5 under extremes. Calculation of pressure drop is quite messy and it is usually sufficient to know the maximum pressure drop, ΔP_M, at the maximum allowable velocity, V_M. Then pressure drop can be estimated for any actual velocity, V_a, by

$$\Delta P = \Delta P_M \left(\frac{V_a}{V_M}\right)^2$$

1.7 ADDITIONAL SUGGESTED READING

Given that there are nearly 500 air pollution control companies in existence in the United States, new developments occur frequently. Many of these developments are more claim than substance; therefore, those people interested in the application of wet scrubbers need to keep abreast of the technical literature being offered, rather than the sales literature.

---

*Registered trademark of E. I. du Pont de Nemours and Company, Wilmington, DE.

TABLE 1.10. Wire Mesh Eliminators Comparison Chart.

| Mesh Styles | | | | UOP Type | Stone & Webster | Mesh Density, lb/ft³ | Mesh Surface Area, ft²/ft³ | Mesh Voids, % |
KOCH	York	ACS	Vico Tex					
4120	421	4BA	380	C	C	12.0	115	97.6
4210	371					10.8	110	97.7
3710						10.0	163	94.0
4310	431	4CA	280	A	B	9.0	86	98.2
3260	326	3BF	415		K	8.0	140	98.4
6440	644					7.3	65	98.5
5310	531	5CA				7.0	65	98.6
9310	931	7CA	160	B	A	5.0	48	99.0
5520		X200	611		L	20.0	450	96.0
5540	333	X100	800		P	27.0	610	94.6
2212	221	8T				4.0	125	97.0
2414	241	8P				4.0	150	97.0

Some additional technical sources to consider adding to your library are listed below. They will provide a thorough reference source for the application of wet scrubbers to environmental control problems.

(1) *Journal of the Air and Waste Management Association*
P.O. Box 2861
Pittsburgh, Pennsyvlania 15230

(2) *Pollution Engineering Magazine*
1350 E. Toughy Ave.
Des Plaines, IL 60017-5080

(3) *Power*
P.O. Box 430
Hightstown, New Jersey 08520

(4) *Environmental Science and Technology*
American Chemical Society
1155 16th Street NW
Washington, DC 20036

(5) Technical Association of Pulp and Paper Industry (TAPPI)
One Dunwoody Park
Atlanta, Georgia 30338

(6) Publications List from:
Institute of Clean Air Companies
700 North Fairfax Street, Suite 304
Alexandria, Virginia 22314

(7) Pollution Equipment News "Catalog and Buyers Guide"
8650 Babcock Boulevard
Pittsburgh, Pennsylvania 15237

(8) Bete Fog Nozzle Catalog
306 Wells Street
Greenfield, Massachusetts 01301

(9) Dwyer Instrument Catalog
Dwyer Instrument Co.
Box 373
Michigan City, Indiana 46360

(10) Crane Co. Catalog (Engineering Section)
Local Representative or
Crane Co.
300 Park Avenue
New York, New York 10022

(11) *Air Pollution Control—Traditional & Hazardous Pollutants, 2nd Ed.*
Howard Hesketh
Technomic Publishing Co., Inc.
851 New Holland Avenue
Box 3535
Lancaster, Pennsylvania 17604

(12) Huntington Alloys Corrosion Chart (Nickel Alloys)
Local Representative or
Huntington Alloys, Inc.
Huntington, West Virginia 25720

(13) Corrosion Charts, Stainless Steel
International Nickel Co.
One New York Plaza
New York, New York 10004

(14) Derakane Chemical Resistance Table
Dow Chemical Co.
Designed Products Dept.
Midland, Michigan 48640

(15) ATLAC Guide to Corrosion Control
I.C.I. United States
Specialty Chemicals Division
Wilmington, Delaware 19897

(16) Stainless Steel in Gas Scrubbers
Committee of Stainless Steel Producers
American Iron and Steel Institute
1000 16th Street NW
Washington, DC 20036

(17) The McIlvaine Scrubber Manual
The McIlvaine Co.
2970 Maria Avenue
Northbrook, Illinois 60062

1.8 REFERENCES

1. *Psychrometric Charts* (Dover, NH: Zimmerman and Lavine, Industrial Research Services, Inc., 1964).
2. *Industrial Ventilation,* 11th ed. (Lansing, MI: Committee on Industrial Ventilation, 1970).
3. Hesketh, H. E. *Air Pollution Control* (Ann Arbor, MI: Ann Arbor Science Publishers, Inc., 1980).
4. *Fan Engineering* (Buffalo, NY: Buffalo Forge Company, 1970).

5. Cheremisinoff, P. and R. Young. *Pollution Engineering Practice Handbook* (Ann Arbor, MI: Ann Arbor Science Publishers, Inc., 1975).

6. Walker, R. P. *Pump Selection* (Ann Arbor, MI: Ann Arbor Science Publishers, Inc., 1972).

7. Hesketh, H. E. *Chem. Engr. Progress* (Vol. 91, No. 10, pp. 98–100, October 1995).

CHAPTER 2

Fiberbed Filters and Wet Scrubbing**

A fiberbed filter is a wet-type collector that uses a Brownian motion capture mechanism to remove submicron liquid and solid particulate aerosols from a gas stream. Traditionally used for acid aerosol collection, fiberbeds have been used in recent years as a highly effective supplement to other wet scrubbing devices.

Fiberbeds bridge the "applications gap" between wet scrubbers and dry filters. They couple wet scrubbing capture techniques with filtration techniques. The fiberbed surface, however, is typically a continually wetted one. In this manner, it can be related to other wet-type collectors.

CECO filters are used for the filtration of very small liquid particles from an air stream or other gas stream. These particles will normally vary in diameter from 100 μ down to perhaps 0.1 μ. A micron is .001 mm; therefore, 25,400 μ is equal to 1 in. Solid particles can have almost any shape: spherical, cylindrical, cubical, rectangular, and irregular. Liquid particles, however, which are CECO's main interest, can be assumed to be spherical.

Particulate systems rarely consist of particles of a given size. Rather, a range of particle sizes is usually encountered. Therefore, if one desires to specifically describe a particulate system, one must define the relative amounts of each size of particulate. This is known as particle size distribution, and it represents a measure of the uniformity of the system from a size standpoint. The density of the particles may be of interest for particles in

---

**This chapter on the fundamentals of fiberbed design and application was written by Mr. Norman Handman, former Chief Engineer of CECO Filters, Inc. (Conshohocken, PA). It provides an excellent overview of the theory of operation of a fiberbed, its design history, and some of its applications. In addition, your attention is directed to the chapter on "Hybrid Scrubbers," which describes some interesting uses of fiberbed technology when coupled to wet scrubbing systems.

the upper range of sizes, where particle inertia and settling velocity may be of importance, but the density becomes of much less interest as the particles become smaller.

Particle size distribution can be represented in several different ways. The most common method is the Frequency Distribution Curve in which the number, or weight of particles in a given size range, is plotted against the average particle size of the range.

In Figure 2.1, Curve A represents a typical distribution curve for systems having a relatively uniform distribution. Curve B indicates a system that might have a skewed distribution, and Curve C shows a system that has more than one peak size. All three systems are relatively common.

Figure 2.2 shows the cumulative curve, in which the fraction of the total number of particles, greater than or less than a given size is plotted against the size.

A third method of describing particle size distribution is by means of a log probability plot. The number or weight frequency is plotted against the logarithm of the particle diameter. If a particulate material follows a log probability frequency distribution, it will plot as a straight line on the graph. If the particulate does not follow a log probability frequency distribution, it will plot as a curve on such a graph. Generally, however, the curvature is not very great.

Particle size distribution information may be represented in any of these three methods and/or in tabular form.

Most of the methods used to determine particle size do not, in fact, actually measure particle size. They usually measure other properties of the

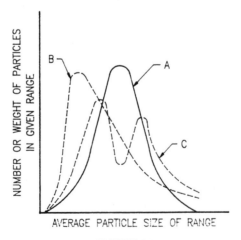

FIGURE 2.1.

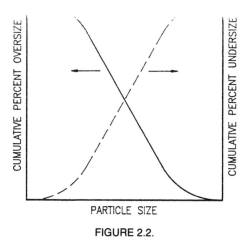

FIGURE 2.2.

particulate, which are then converted to an apparent or effective diameter by means of analytical or empirical relationships that may or may not be totally correct. The method by which the particle sizes were determined should be considered before making any commitments.

Fine particles, generally defined as less than 3 μ in diameter, are one of the most important forms of air pollution. First, these fine particles can remain airborne for extended periods. Secondly, their greater ability to obstruct and scatter light causes the smog and haze problem. And, most important, they are a health problem since, in contrast to larger particles, they can bypass the body's respiratory filters and penetrate deep into the lungs. The chemical and physical characteristics of fine particulate can further aggravate their health impact. Because of their high surface area, some fine particulates have been identified as transport vehicles for gaseous pollutants (both absorbed and reacted) and can, therefore, produce synergistic effects harmful to health. Many fine particles are metallic and, hence, are chemically and catalytically active.

Some processes emit solid and liquid hydrocarbons. Some such processes are lubricating oil emissions from gas turbines, asphalt storage tank emissions, scrubber emissions, solvent emissions from printing plants, etc. In general, these hydrocarbon aerosols are generally in the range of 0.2–3 μ in diameter.

The catalytic oxidation method of manufacturing sulfuric acid produces extremely noxious aerosols of sulfur trioxide, SO_3, and sulfuric acid, H_2SO_4. The particle size distribution of the aerosol varies somewhat with the product being manufactured and the particular piece of equipment. For example, the particle size distribution from the drying tower, the primary absorbing tower, the secondary absorbing tower, and the ammonia scrubber may differ from one another. The particle size distribution will also vary with the strength of acid being produced. In general, oleum plants will produce aerosols of about 0.6 μ. In non-oleum producing plants the typical aerosol is 1–2 μ. All these plants can produce aerosols ranging from 0.2–10 μ. The quantity of mist generated varies with the fraction of design capacity at which the plant is operating, being highest at high capacity.

The quantity and size of aerosol also varies with the raw material. A plant burning dry, high-grade sulfur will produce less mist than a plant burning organic sulfates or spent acid. Typically, before the advent of fiber mist eliminators, sulfuric acid plants were characterized by the thick white acid plumes emanating from their stacks. Clearly the health problems of sulfuric acid mist are a matter of serious concern. CECO filters are presently in use in a number of acid plants and have resulted in a most satisfactory solution to the problem.

The transfer unit, N_t, is a convenient and common way to express filter efficiency:

$$N_t = \ln \left[\frac{1}{1 - \text{eff}} \right] \tag{1}$$

It is analogous to the transfer unit often used in describing the action of absorption towers. Corresponding values of N_t and efficiency are listed in Table 2.1.

Figure 2.3 illustrates the variation of the transfer unit with the face velocity for various values of fiber diameter and bulk densities. Curve C repre-

TABLE 2.1.

Transfer Unit (N_t)	Efficiency (eff)
.1	9.5
.5	39.4
1	63.2
2	86.47
3	95.02
4	98.17
5	99.4
6	99.75
7	99.91
8	99.97
9	99.9877
10	99.9955

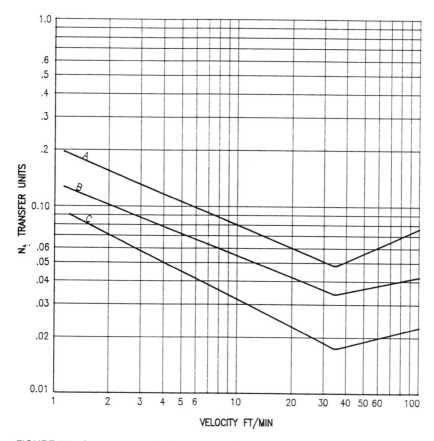

FIGURE 2.3. Curves A, B and C illustrate the efficiencies of different diameters and bulk densities versus superficial air velocity.

sents course fiber diameter and low bulk density. Curve A is for a filter of fine fiber and high bulk density. Curve B is for a filter of intermediate fiber diameter and bulk density.

2.1 COLLECTION BY DIRECT INTERCEPTION

A very small particle having essentially no mass will have no inertia and its center will follow the air stream lines as they flow around the fiber. If the particle has a finite diameter, d_p, it will touch the collecting fiber when its center approaches within a distance of $d_p/2$ of the fiber surface. This effect is called flow line or direct interception and is illustrated in Figure 2.4.

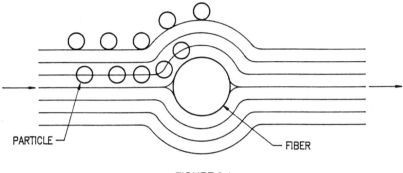

FIGURE 2.4.

2.2 COLLECTION BY INERTIAL INTERCEPTION

A larger particle, having an appreciable mass, moving toward a fiber which is perpendicular to the direction of air flow will not move parallel to the fluid stream lines. Because it has mass, and therefore momentum, it will tend to move along its original path.

Its momentum will make it less subject to deviation from its course when the air stream lines spread sideways around the collecting fiber, as illustrated in Figure 2.5.

2.3 COLLECTION BY BROWNIAN MOTION OF PARTICLES

This phenomenon is also called Brownian Motion Diffusion of particles. It is the principal mechanism for the collection of small, submicron par-

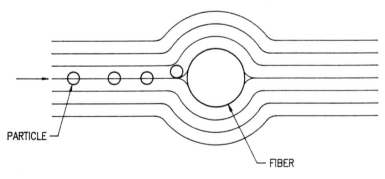

FIGURE 2.5.

ticles in CECO filters. Aerosol particles exhibit a significant random motion caused by collision of air molecules with the particles. Albert Einstein, in 1905, developed Equation (2), which describes this random motion.

$$\Delta s = \sqrt{\frac{4RTC_c t}{3\pi^2 \mu N d_p}} \qquad (2)$$

where

Δs = distance in feet moved by a particle in time, t
R = universal gas constant = 1545 for air
T = absolute temperature, °Rankine
C_c = Cunningham correction factor
μ = air viscosity, lb/ft/sec = 12.1 × 10^{-6} lb/ft/sec
N = Avogadro's constant = 2.76 × 10^{26} molecules per pound mole
d_p = particle diameter, feet
t = time, seconds

Particles will diffuse or migrate from an area of high concentration to an area of low concentration in a manner analogous to the flow of a gas from an area of high to low concentration. The particles will collect on the filter fibers and the air in contact with the fibers will be at zero concentration. Particles will therefore migrate from the main air stream to the filter fibers, where they will coalesce and drain off.

The coefficient of diffusion, also derived by Einstein, is:

$$D_v = \frac{RTC_c}{3\pi\mu N d_p} \qquad (3)$$

where

D_v = diffusion coefficient
μ = gas viscosity

In Table 2.2, the variation of Cunningham Correction Factor, C_c, and displacement, Δs, is indicated, with particle diameter due to Brownian Motion. Smaller particles move further in a given time, hence, are more likely to hit a filter fiber and be filtered out of the air stream.

An approach to presenting aerosol collection efficiencies was developed at the Hanford Atomic Bomb Works by the Atomic Energy Commission. It

TABLE 2.2.

Particle Diameter in Microns	Cunningham Correction Factor, C_c	Displacement in 1 sec in Microns
0.1	2.88	29.4
0.25	1.682	14.2
0.5	1.325	8.92
1.0	1.165	5.91
2.5	1.064	3.58
5.0	1.032	2.49
10.0	1.016	1.75

was found experimentally that for submicron particles, collection efficiency of a given filter could be expressed as:

$$\text{eff} = 1 - 10^{-DF} \tag{4}$$

DF is decontamination factor.

The decontamination factor, in turn, can be determined from Equation (4), from a filter's physical dimensions, where

$$DF = \frac{CL^a B^b}{V^c} \tag{5}$$

where

L = filter thickness, inches
B = bulk density, pounds per cubic foot
V = air velocity, feet per minute

C, a, b and c are constants that must be determined experimentally for each fiber diameter.

Figure 2.6 is typical for a filter made of a particular fiber diameter. It plots the decontamination factor versus the superficial velocity for a family of filters of the same fiber diameter and increasing bulk densities. It indicates that the efficiency increases with increasing bulk density. Table 2.3 shows the value of efficiency for various values of the decontamination factor.

If one experimentally determines the performance of a particular fiber so that the constants C, a, b and c are determined, it is then possible to predict the performance of that fiber at other thicknesses, bulk densities, and velocities.

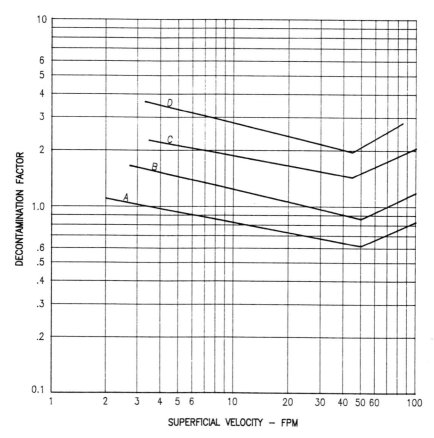

FIGURE 2.6. Decontamination factor versus velocity for constant fiber diameter and varying bulk densities. All four curves are for the same fiber diameter. Curve "A" is the lowest and Curve "D" is the highest bulk density.

TABLE 2.3.

Decontamination Factor	Efficiency
.1	.21
.5	.68
.8	.84
.9	.87
1.0	.9
1.5	.97
2.0	.99
2.5	.997
3.0	.999
3.5	.9997
4.0	.9999
4.5	.99997
5.0	.99999

2.4 PRESSURE DROP

The pressure drop across a clean filter is a function of bed thickness, air velocity, fiber diameter, bulk density, air density and air viscosity. A generalized equation that describes pressure drop is:

$$\Delta p = 8C_f \mu V_0 L_f (1 - \epsilon) D_b^2 \tag{6}$$

where

C_f = constant that must be experimentally determined—it is a function of ϵ and N_{Re}

μ = air viscosity

V = air velocity

L_f = bed thickness, inches

ϵ = void fraction of bed (bulk density)

D_b = fiber diameter

Figures 2.7, 2.8 and 2.9 illustrate the variation in pressure drop versus velocity for constant fiber diameter and varying bulk densities. In the Atomic Energy Commission report previously referred to herein, Blasewitz et al. [2] presented the following equation for pressure drop:

$$\Delta P = KL^x \varrho^y V^z \tag{7}$$

where

P = pressure drop, inches of water
K = constant for each fiber to determined experimentally
ϱ = bulk density

Exponent x equals 1, indicating that pressure drop is linear with air velocity. This equation is for air at normal temperature and pressure. Corrections must be applied for air at higher or lower temperatures and/or pressures. Exponent y is generally equal to 1.5 to 1.6 and exponent z equals 1, indicating that pressure drop is linear with velocity.

In general, the smaller the fiber diameter, the higher the pressure drop.

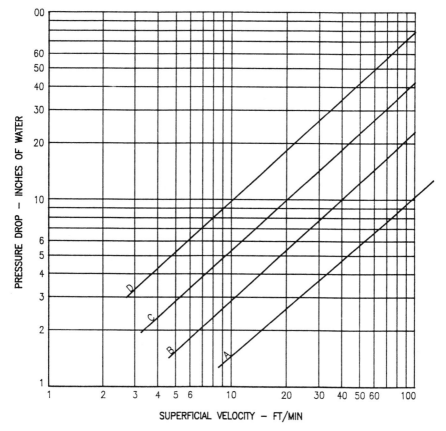

FIGURE 2.7. Velocity versus pressure drop for constant fiber diameter and varying bulk densities. All curves are for fibers of equal diameters. Curve "A" is the lowest and Curve "D" is the highest bulk density.

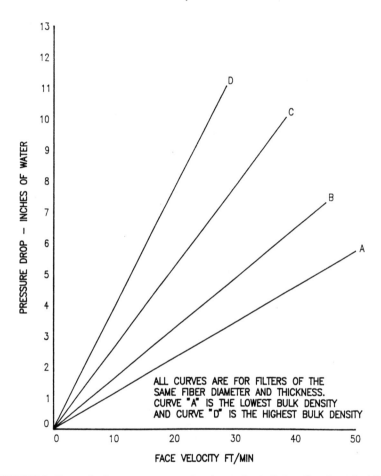

FIGURE 2.8. Face velocity versus pressure drop for varying bulk densities at constant fiber diameter and filter thickness. All curves are for filters of the same fiber diameter and thickness. Curve "A" is the lowest bulk density and Curve "D" is the highest bulk density.

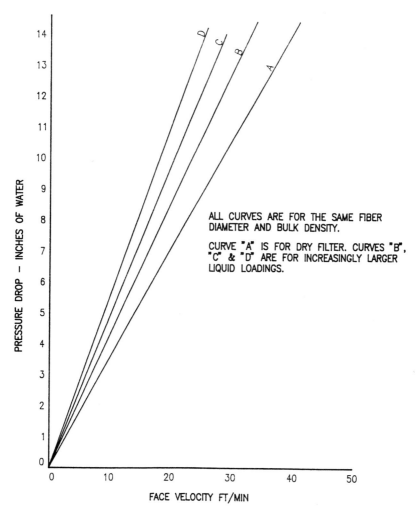

FIGURE 2.9. All curves are for the same fiber diameter and bulk density. Curve "A" is for dry filter, Curves "B," "C" and "D" are for increasingly larger liquid loadings.

A family of curves such as Figures 2.7, 2.8, and 2.9 can be drawn for each fiber at various velocities and bulk densities. The equations and the curves are for clean filters. As aerosol is collected on the filter, the pressure drop will increase. This increase in pressure is a function of the quantity of aerosol mist in the air stream, density of the liquid, viscosity of the liquid, and surface tension of the liquid.

2.5 REENTRAINMENT

Air flowing at a velocity of 1200 ft/min or higher across a wet filter can reentrain the liquid droplets on the surface of the filter. This is a distinct possibility in the case of air flowing from outside to inside of a long filter. All the air must eventually exit from the mouth of the filter. Assume, for example, a 16 in. × 12 in. filter. The area of the mouth, or opening of the filter is:

$$\frac{\pi}{4}(1)^2 = .7854 \text{ ft}^2$$

$$.7854 \times 1200 = 942 \text{ cfm}$$

This is the maximum air quantity that a 16 in. × 12 in. filter can handle in outside to inside flow. If the velocity of flow across the filter is 20 ft/min, for example, the filter inner surface area is $942 \div 20 = 47 \text{ ft}^2$. The maximum length of the filter then is:

$$\frac{47 \text{ ft}^2}{\pi \times 1} = 14.96 = 15 \text{ ft}$$

The 20 feet per minute velocity might have been determined by pressure drop considerations. Now let us assume a 24 in. × 20 in. filter. The area of the mouth of the filter is:

$$\frac{\pi}{4}\left(\frac{20}{12}\right)^2 = 2.18 \text{ ft}^2$$

$$2.18 \times 1200 = 2618 \text{ cfm}$$

The maximum air flow through a 24 in. × 20 in. filter in outside to inside configuration is 2618 cfm. If the filter is selected at a face velocity of 20 ft/min, the area of the filter is 2618 ÷ 20 = 131 ft². With:

$$131 \text{ ft}^2 = \pi \left(\frac{20}{12}\right) L$$

Then the maximum length, L, of the filter is 25 ft. There is usually no problem of reentrainment from filters when the air flow is from inside to outside of the filter.

Let us assume an air stream of 10,000 scfm at 300°F and 125°F wet bulb temperature. Assume that the air contains 7500 parts per million of dibutylphthallate, a common plasticizer often used in vinyl extrusion plants. The plasticizer produces a noxious blue-grey aerosol. How is the hydrocarbon removed from the air stream?

Reference to Lange's *Handbook of Chemistry* gives the following equation:

$$\log VP = A - \frac{B}{T + C} \tag{8}$$

where

VP = vapor pressure in millimeters of mercury
A = 6.6398
B = 1744.2
C = 113.69
T = temperature in °Centigrade

Equation (8) now becomes:

$$\log VP = 6.6398 - \frac{1744.2}{T + 113.69} \tag{9}$$

Substituting values of T, in °C, and solving for log of VP, and then taking the antilog of the log of VP to arrive at the vapor pressure, the following results are obtained in Table 2.4.

Degrees centigrade can be converted to the reciprocal of degrees Rankine as follows:

$$°F = 1.8 \times °C + 32$$

$$°R = °F + 460$$

$$1/°R = \text{Reciprocal}$$

TABLE 2.4.

T (°C)	log VP	VP (mmHg)	$1/T$ (1/°R)
40	−4.71	1.95×10^{-5}	.00177
50	−4.01	9.77×10^{-5}	.00171
60	−3.40	3.96×10^{-5}	.00166
70	−2.86	.0013	.00162
80	−2.36	.0043	.00157
90	−1.92	.0119	.00153
100	−1.52	.0300	.00149
150	.025	1.059	.00131

These corresponding values are shown in column 4 of Table 2.4. We can then chart $1/°R$ versus the log VP, as shown in the graph of Figure 2.10.

Assume that the leaving air cannot contain more than 50 parts per million of organic plasticizer. Then the leaving vapor pressure must be lower than

$$\frac{50}{1,000,000} \times 760 = 0.038 \text{ mmHg}$$

$$\log 0.038 = -1.42$$

From the graph, when log $VP = -1.42$, can find $1/°R = .001485$, and can calculate $R = 673°$ and $F = 213°F$. Therefore, the air must be cooled from 300°F to 213°F. Actually, the lower we reduce the temperature, the less plasticizer will remain in the air.

The amount of heat that must be removed in order to reduce this temperature is:

$$H = 10,000 \times 1.08(300 - 213)$$

$$= 939,600 \text{ Btu per hr}$$

At the reduced temperature, the plasticizer will condense to an aerosol and will be filtered out by the CECO filters.

This amount of heat can be removed in two ways. The first and least expensive way is to spray cold water into the air stream. The water will evaporate into the air stream. Each pound of water that evaporates will absorb approximately 1000 Btu. Consequently, 939,600 = 1000W, where W is the pounds of water that will be evaporated per hour.

$$W = 939.6 \text{ lb/hr}$$

$$\frac{(939.6 \text{ lb/hr})}{(8.33 \text{ lb/gal})(60 \text{ min/hr})} = 1.88 \text{ gal/min evaporated}$$

$$10,000 \text{ scfm} = 45,000 \text{ lb of air per hour}$$

Air at 300°F and 125°F wet bulb contains 0.05 lb of water vapor per pound of dry air. When the air is cooled by the evaporation of water

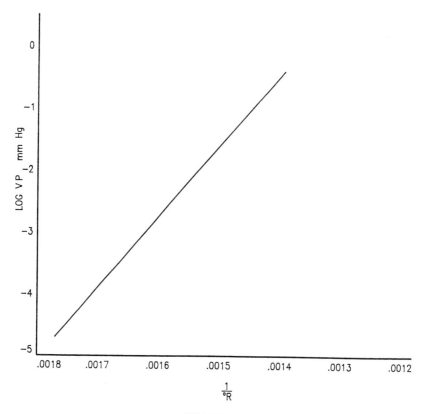

FIGURE 2.10.

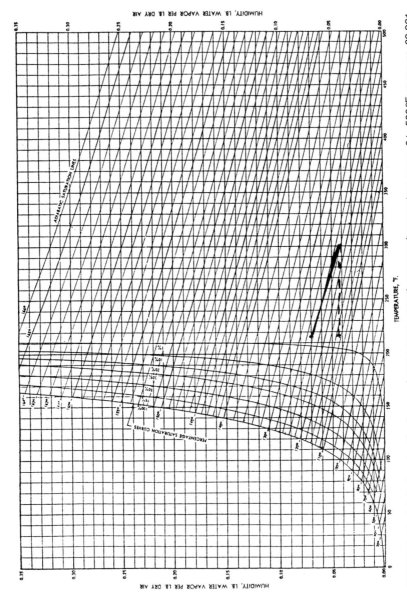

FIGURE 2.11. Adiabatic saturation lines and percentage saturation curves (temperature range, 0 to 500°F; pressure, 29.921 in. Hg).

84

(adiabatic cooling) the air increases in moisture content along its wet bulb temperature as it cools.

$$\frac{939.6 \text{ lb of water/hour}}{45,000 \text{ lb of air/hour}} = \frac{.021 \text{ lb of water added}}{\text{lb of dry air}}$$

$$.021 + .05 = .071 \text{ lb of water per lb of dry air}$$

As shown, there would be 0.071 lb of moisture per pound of dry air at 213°F (see psychrometric chart in Figure 2.11). This corresponds to a dew-point of 116°F. In cold weather, this moisture will condense when it comes in contact with the atmosphere. It can create a troublesome plume in conjunction with the residual 50 parts per million of plasticizer in the air stream.

A better but more expensive way to cool the air is to cool it by means of a water cooled heat exchanger. In this case, the air is cooled from 300°F to 213°F along a line of constant water vapor dewpoint. There is no increase in the water vapor content of the air as it cools. The plasticizer condenses out and is removed by the CECO filters. The water quantity is determined from the cooling load and the water temperature rise in the coil. In this case, the final dewpoint is the same as the initial dewpoint, approximately 105°F. This air, too, can condense out in cold weather but is much less likely to do so.

When cooling with water sprays the air can be cooled and dehumidified if there is sufficient water at low temperature. The water temperature will rise and the air temperature and water vapor content of the air will be reduced. If the water is merely recirculated through the air stream it will come to the air wet bulb temperature and the air will cool and humidify along its wet bulb line.

2.6 REFERENCES

1. Blasewitz and Judson, "Filtration of Radioactive Particles by Glass Fibers," *Chemical Engineering Progress,* January 1955.

2. Blasewitz, Carlisle, Judson, Katzer, Kurtz, Schmidt, and Weidenbaum, *Filtration of Radioactive Aerosols by Glass Fibers,* Handford Works, Richland, Washington Energy Commission, Atomic Energy Commission, General Electric Company, April 16, 1951.

3. Brink, J. A., Jr., "Acid Mist Control with Fiber Mist Eliminators," *Canadian Journal of Chemical Engineering,* June 1963.

4. Brink, Burggrabe, and Greenwell, "Mist Eliminators for Sulfuric Acid Plants," *Chemical Engineering Progress,* November 1968.

5. Duros and Kennedy, "Acid Mist Control," *Chemical Engineering Progress,* September 1978.

6. *EPA Fine Particle Scrubber Symposium,* National Environmental Research Center, NIT, A.P.T. Inc., October 1974.

7. Fairs, "High Efficiency Fiber Filter for the Treatment of Fine Mists," *Transactions Institute for Chemical Engineers,* Vol. 36, 1950.

8. Lapple, C. E., *Fibrous Aerosol Filters,* Chemical Engineering Department, Ohio State University, AEC Los Alamos Meeting, September 1953.

9. Lapple, C. E., "Mist and Duct Collection," *Heating, Piping, and Air Conditioning,* February 1946.

10. Ramskill and Anderson, *The Inertial Mechanism in the Mechanical Filtration of Aerosols,* Naval Research Laboratory, Washington, DC, 1957.

11. Ranz, W. E., "The Role of Particle Diffusion and Interception in Aerosol Filtration," *Engineering Experiment Station Technical Report,* University of Illinois, September 1, 1952.

12. Ranz, W. E., "The Impaction of Aerosol Particles on Cylindrical & Spherical Collectors," *Engineering Experiment Station Technical Report,* No. 3, University of Illinois, March 31, 1951.

13. Stairmand, C. J., *Dust Collection by Impingement and Diffusion,* Institute of Chemical Engineering, 1950.

14. Strauss, Werner, *Air Pollution Control,* John Wiley & Sons, Inc., 1971.

15. Thomas and Lapple, "Deposition of Aerosol Particles," *A.I. Ch.E. Journal,* Volume 7, No. 2, page 203.

16. Thomas, D., "Deposition of Aerosol Particles in Fibrous Packaging," Ph.D. Thesis, Ohio State University, 1953.

17. Wong, J. B., N. F. Johnstone, "Collection of Aerosols by Fiber Mats," *Technical Report,* No. 11, U.S. Atomic Energy Commission, October 31, 1953.

18. Wong, Ranz, and Johnstone, "Inertial Impaction of Aerosol Particles on Cylinders," *Journal of Applied Physics,* Volume 26, No. 2, February 1955.

19. Wright, Stasay, and Lapple, "High Velocity Air Filters," *WADC Technical Report 55-457,* Wright Air Development Center, U.S. Air Force, October 1951.

CHAPTER 3

Scrubber Applications

3.1 INTRODUCTION

A number of typical scrubber applications are presented in this chapter to give the reader specific data for those presented. General information relative to scrubber applications can be extrapolated to similar industrial situations as appropriate. Each section includes a discussion and an "application guide" table at the conclusion of the section. Table 3.1 is an application guide with explanatory notes.

3.2 ASPHALT PLANT SCRUBBERS

Many owners of hot mix asphalt plants responded to the requirements of the 1970 Clean Air Act and its state derivatives by installing wet scrubbers. These devices, ranging from spray chambers to cyclonic scrubbers to venturi scrubbers, saw application on both portable and fixed asphalt plants.

The emissions from an aggregate dryer include dusts of less than 1 μ (usually less than 10% is of this size), with most of the dust falling in the range of 10–80 μ (70–80% of the dust is in this range). Thus the problem is one of removal of predominantly large particulate with a naggingly significant quantity of fines. We therefore see such a variety of scrubbers (some homebuilt) in this industry.

It is quite simple to remove the large particulate. Simple spray chambers using full cone nozzles are able to remove over 95% of the incoming 10–80 μ dust. Many dryers are able to reach code levels when the infeed produces only a small (less than 3–4%) quantity of fines. When this quantity rises, the operator is suddenly out of compliance. His simple spray chamber can-

TABLE 3.1. Example Application Guide.

Type of Scrubber: General wet scrubber type typically employed. Throat: If venturi scrubber, type of throat (annular, rectangular, etc.). Actuation: For adjustable throats, manual or automatic. Materials: Typical construction materials. Access Ports: Location and type recommended. Pressure Drop: Common pressure drop. Outlet Grain Loading: Loading obtainable at listed pressure drop. Inlet Temperature: Common inlet temperature. Outlet Temperature: Common outlet temperature. L/G: Common liquid-to-gas ratio, gal/1000 ft^3 of gas. Droplet Control: Separators, wear plates, etc. Separator Type: Cyclonic, mesh pad, combination, vane, etc. Inlet Velocity: Typical inlet velocities. Outlet Velocity: Common outlet velocities. Test Method: If special testing is required, list here. Special Considerations: Any special requirements of this application.

not remove these particulates at anywhere above 30–40%. Wetting agents may be tried, higher liquid-to-gas ratios may be used, but most modifications are unsuccessful.

We then find the cyclonic scrubber being used to remove greater quantities of these particulates. Unfortunately, these inertial devices do not improve the removal efficiency to the 50–70% required on the fines. When they are operated at near 10 in. w.c. the efficiency is met or exceeded; however, the units can quickly become entrainment devices since the tangential velocities are over 120 ft/sec. Again, the predominant particulate is removed; it is the nagging residual flow of fines that prompts the added energy consumption.

Thus, venturi scrubbers are often employed. One manufacturer, PolyCon Corporation (now owned by the Environmental Elements Division of Koppers), sold over 60 of these units for asphalt plant applications in the years 1968–1976 (Figure 3.1 shows this system). The venturi has an open header gravity-type hopper approach of rectangular configuration so as to suit the typically rectangular breeching found in these dryers. The units use no spray nozzles which might plug, instead they rely on a simple rectangular venturi (sometimes adjustable, especially for draft-sensitive oil firing) and a flooded elbow for abrasion-resistance, leading to a conventional cyclonic separator. Some units are horizontal, others vertical. Each uses approximately 6–8 gallons per minute (gpm) of water recycled to a settling pond, and has venturi draft losses of 10–18 in. w.c. and separator losses of 2–3 in.

w.c., including inlet and outlet connections. Figure 3.2 gives efficiency information.

Mild steel can be used, with 1/4-in. plate being selected for the throat and 3/16-in. for other sections. Some cyclonic separators included replaceable wear liners, with venturi throats being flanged for replacement. These items were the high-wear sections.

Specific details of design include antiswirl baffles in the separator sump to prevent cavitation, fans on the clean air side to reduce fan wear from the abrasives encountered, open-impeller (sometimes rubber-lined) centrifugal pumps for recycle, and draft or pressure drop indicators on the venturi throat. Compact systems placed the fan before the scrubber but after a product recovery cyclone (either a cast tubular multicyclone or grouped conven-

FIGURE 3.1. Wet scrubber on asphalt plant application (courtesy Environmental Elements Division of Koppers).

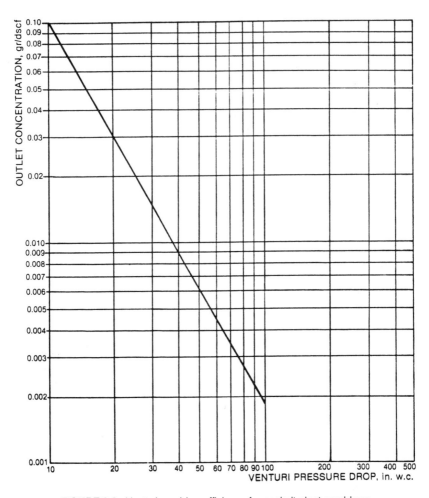

FIGURE 3.2. Venturi scrubber efficiency for asphalt plant scrubbers.

tional cyclones) so that the stack could be placed on top of the scrubber. Fans were straight radial-bladed centrifugals, of paddle wheel design, with split housings for wheel removal, sprays for cleanout, and drains.

The hallmark of designs like these is the ability to run unattended, with longest possible periods between repair procedures even though the unit is subjected to extremely demanding service. Inlet grain loadings on these applications can exceed 10 gr/scf of abrasive dust. The scrubber recirculates scrubbing liquid in an effort to reduce water loss. This results in solids concentrations of up to 10% (wt) solids. Therefore, all headers are open, with minimum numbers of bends or dead ends. Some operators use rubber hose connections on the liquid circuit so as to reduce maintenance cost and abrasive wear. (The best hose includes a natural rubber liner for abrasion resistance.)

Drains are sized for 2–4 ft/sec liquid velocities with adequate access provided by at least one 18-in. manhole in the separator. Separators may be made from old cyclonic air washers if the plant is so equipped. Vertical velocities of less than 10 ft/sec must be possible, however, if a conversion is to be contemplated.

If a vertical stack is to be used on top of the separator, outlet velocities of less than 30 ft/sec are recommended. Higher velocities will likely cause carryover.

If the scrubbing liquid is sent to a pond, the return line must be maintained as far from the drain line as possible so as to prevent the inadvertent return of unsettled solids. Some installations with a space problem use a U-shaped pond (with a dike in the center) so that a long liquid settling path is produced; yet the return and drain pipes are physically close together to reduce pipe runs, as shown in Figure 3.3.

Pumps are usually sized for up to 12 gal/1000 acf so that the pump alone may be used for pressure drop control on venturi systems over 10 in. w.c. This provides for a 10–15% pressure drop adjustment through throat area reduction caused by displacement. The flow rate of the water is calculated in cubic feet per second (cfs) and then divided by the velocity in the throat. This gives the area displaced by the water. Subtracting this figure from the

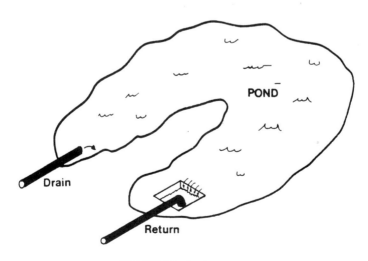

FIGURE 3.3. Spoils pond.

total area of the throat produces the net *effective* area. One then compares this velocity with the velocity required for a given pressure drop, and through iteration, the new pressure drop is determined. Please note that the water velocity and air velocity will be slightly different, because an increase in the gas velocity will not increase the water velocity significantly since it is incompressible.

Other scrubbers use adjustable damper blade throats which provide for velocity changes and therefore pressure drop adjustment. An additional advantage is that they may be closed during fan startup to reduce the draft on the burner. For oil-fired dryers, this feature is particularly important.

The throat is typically manually adjusted by an external lever or wheel. A locking pin or bolt holds the throat in its proper position. Automatic throats use a pressure drop cell which senses the draft on the dryer vs. atmosphere or the venturi pressure drop, translating this information into a milliamp or millivolt signal which actuates a power positioner mounted on the throat damper shaft. A manual override is necessary for burner start. Typically, the positioner is a spring return which fails in the open position so that the gases may be ventilated through the system.

Minimum instrumentation includes pressure drop indicators on the scrubber, pressure gauges with snubbers on the liquid headers (especially on those headers including spray nozzles), flow switches on the drain line (to indicate flow loss, therefore pluggage or pump failure), and outlet temperature indicators (dial-type thermometers). When the outlet temperature rises noticeably, the operator knows that the venturi or cyclonic scrubber is not saturating the gas stream properly.

Horizontal venturi scrubbers can pose some difficult liquid distribution problems. Many are spray approach types that must have a constant, uniform spray pattern. A single bulk spray nozzle of full cone design can sometimes improve the operation of horizontal venturis. The throat should have a width not exceeding 1/4 of the length, forming a slot rather than a rectangular box. This reduces the pattern of flow to the bottom of the throat. The long dimension of the throat should be vertical so as to provide a minimum ramp angle of the approach to the throat. Many improperly operating horizontal venturis can be corrected by careful modification.

An excellent reference for the applications engineer's file is the National Asphalt Paving Association pamphlet [1]. Table 3.2 presents an application guide for asphalt plant scrubbers.

3.3 BARK BOILER SCRUBBERS

Boilers that are fired by bark and wood shavings are generally referred to as bark boilers. Those that fire a variety of wood waste, including bark, are

TABLE 3.2. Asphalt Plant Scrubbers: Rock Dryer Application Guide.

Type of Scrubber: Venturi with flooded elbow, cyclonic separator.
Throat: Adjustable, rectangular, replaceable throat blade.
Actuation: Manual or automatic on draft control, flame out open.
Materials: Mild steel, 0.25-in. plate.
Access Ports: In venturi elbow and separator.
Pressure Drop: See graph in this section, 10–18 in. w.c.
Outlet Grain Loading: See graph, typically 0.05 gr/dscf.
Inlet Temperature: 250–500°F.
Outlet Temperature: 135–150°F (lower if recycled from pond).
L/G: 10–12 gal/scf.

Droplet Control: Cyclonic separator with wear plate.
Separator Type: Cyclonic.
Inlet Velocity: 50 ft/sec.
Outlet Velocity: 30 ft/sec with stack; 60 ft/sec to fan.

Test Method: Front catch of Method 5. If oil-fired, adjust burner before test to
  reduce carbon carryover.

called wood waste boilers. Others that fire wood waste along with an auxiliary fossil fuel are called combination fired boilers.

Obviously, there exists a great deal of overlap, depending on the type of
fuel fired. The mixture of fuel is of great importance to the applications engineer.

Important considerations are:

(1) Is the wood cut in sandy locations? Is it dragged on the ground after cut-
ting? Is it hardwood or softwood? Does the mill stack pulpwood or furnish as logs or does it chip and then store?

It has been shown that the silicates in wood bark come from sand particles physically attached to the bark and also from the bark itself. The
silicates are evolved in the boiler, producing sand particles which contribute to the mass emission rate and submicron silicon dioxide which
contributes to a visible emission.

Hardwood tends to have greater quantities of silicates than softwood,
except where the softwood grows in areas of high sand content (the Gulf
Coast area, certain sections of the South near the Mississippi, and
other areas). Many existing boilers utilize multitube collectors, which
effectively remove the high-density silicates. An analysis of the collected ash from this device can help estimate total silica to be expected
in the collector. A stack test will also reveal the percentage of fine particulate encountered.

Low-energy (4–6 in. w.c.) collectors can effectively remove the
silicates and char of 5 μ and greater. The problem occurs with the sub-

micron silicon dioxide fume or the fine oxides introduced through the firing of auxiliary fuel.

If the furnish is stacked and stored prior to debarking, a benefit can sometimes be realized in the reduction of topical sand particles. If it is barked and then chipped or stored without bark, there is typically a higher percentage of sand particles. If the furnish is floated in salt water, the residual chlorides and metallic salts present in the sea water will evidence themselves by corrosion and small particle evolution during combustion. An accurate particle size analysis is therefore mandatory.

(2) Are ends, butts and wood waste also burned? What percentage of the total feed does this represent?

In general, burning the parent wood (cambium) produces char. Burning the bark tends to produce finer particulate. Burning auxiliary fossil fuel such as oil introduces additional submicron particulate and may contribute to increased wet catch (impinger catch) during a stack test in the form of aldehydes and other organics.

When ends, butts, and pieces of compression wood are burned, greater char emissions may occur. Many bark boiler pollution control systems start with a multitubular dry collector for removal of large char and $+5$-μ char, followed by a low- to medium-energy wet collector in order to properly address this two particle problem.

(3) If a multitube collector is used, is the char reinjected? If it is, the percentage of fines to the scrubber will likely increase, requiring higher pressure drops.

If there are any leaks on the multitube hopper lock, the grain loading to the scrubber will increase, requiring a higher efficiency for a given outlet loading. Systems which reinject char to recover its heating value also tend to retain in its closed cycle a higher percentage of fines, which tends to increase wear on any rotary lock which may be fitted.

A recent study by the National Council for Air and Stream Improvement analyzed this problem of char reinjection. The relationship of outlet dust loading to venturi pressure drop was presented in the *McIlvaine Scrubber Manual* [2] as seen in Figures 3.4 and 3.5.

(4) It is evident that four contributing factors combine to produce the net outlet emission:
- percentage of reinjection
- sand content of fuel
- type and quantity of auxiliary fuel
- fuel moisture

Worst case emissions would be a bark boiler with full reinjection burning wet coastal softwood bark in combination with coal. Best case

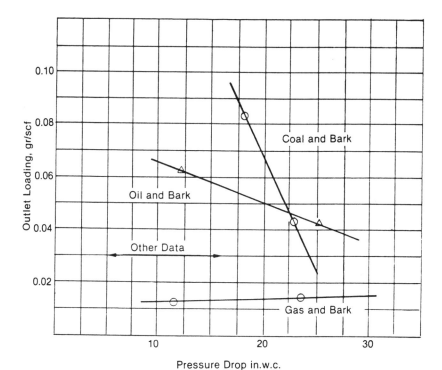

FIGURE 3.4. Efficiency vs. pressure drop for venturis on bark boilers (source: *The McIlvaine Scrubber Manual* [2]).

emissions would involve a gas-fired dry hardwood boiler without reinjection.

(5) Is the mechanical collector functioning properly?

Common problems include hopper recirculation and broken or damaged tubes. Particulate size or loading tests conducted on improperly operating multicyclone collectors are misleading, potentially indicating scrubber pressure drops far below those really required. This occurs because uncollected char and sand pass through the collector, contributing on the outlet side to a higher mass percentage of large particulate, which a low-energy scrubber removes effectively. Control this bypass and the multicyclone removes the large higher-density sand particles, letting the smaller sand and lighter char particles pass through, raising the percentage of fine particulate to be removed by the scrubber.

Hopper bypassing can be controlled by using baffles or building the hoppers in compartments. Broken tubes or tube vanes must be replaced.

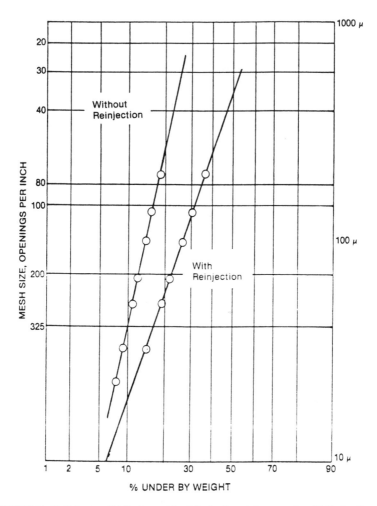

FIGURE 3.5. Particle size distributions of bark boiler fly ash (source: *The McIlvaine Scrubber Manual* [2]).

Multicyclone operation can be judged after a wet scrubber is installed by observing the percentage of "floaters" in the scrubber recycle tank, their increase indicating lost tubes or internal leaks.

Typical wet scrubbers installed on bark, combination, or wood waste boilers include Western Precipitation's Doyle-type impactor scrubber (see Figures 3.6 and 3.7), a variety of cyclonic scrubbers and some low- to medium-energy venturi scrubbers as manufactured by Ducon, Neptune AirPol, and American Air Filter. An AirPol system is shown in Figure 3.8. The venturi scrubbers operate at pressure drops of 4–10 in. w.c., with the low drop ranges normally featuring spray approaches. Venturi scrubbers do not hydraulically atomize well until pressure drops of 6–8 in. w.c. are attained. The Doyle scrubbers operate at 4–6 in. w.c. Efficiencies produced at these drops are above 95% removal. The outlet loadings range from 0.015 to 0.070 gr/scf depending on the type of boiler and mechanical operation of the scrubber.

Venturi scrubber pressure drops of 12–16 in. w.c. produce outlet loadings

FIGURE 3.6. Dual Western Precipitation scrubbers on bark boil flue gas system (courtesy Western Precipitation Division of Joy Manufacturing).

FIGURE 3.7. Grouping of four impactor-type Doyle scrubbers on wood waste boiler (courtesy Western Precipitation Division of Joy Manufacturing).

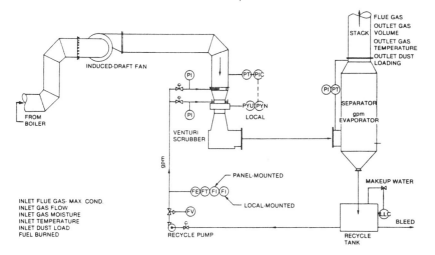

FIGURE 3.8. Process flow diagram for bark boiler scrubber system (courtesy Air Pollution Industries, Inc.).

of less than 0.02 gr/scf when auxiliary oil is fired at up to 30% of the net heat input. Adjustable venturi scrubbers are excellent choices where the boiler is lit off with auxiliary fuel but then shifts back to full wood or bark fuel once the bed is established. With a variable-speed fan, the scrubber can be reduced in pressure drop and the gas volume trimmed to that required for proper combustion of fuel.

Since sand, of specific gravity of 2.3–3.5, and bark of specific gravity of 0.15–0.5 are a combined emission, the water circuit of the scrubber must be carefully designed. One has a problem of "floaters" (char) and "sinkers" (sand). This problem is more pronounced when spray nozzles are used in the system design where plugging may occur. The char particles do not permit stagnant or semistagnant zones in the scrubbers; thus the rule of thumb is to get them out . . . quickly! The scrubber sump is typically marginal in size; the drain is sized for 1–2 ft/sec flow, and an external recycle tank is used for separation of these two particulate types.

A V-notch weir tank is commonly used to skim the floaters from the recycle, the latter being drawn off the tank not at the bottom, but at the side, well above the zone of sand accumulation. The sand is bled out, with makeup losses accommodated through introduction at the gas inlet to the scrubber. The char is sluiced off. Being quite small in size, many mills simply dispose of it. Others reclaim it by sending it over a screen to partially dewater it and then send it onto the bark (fuel) infeed. The latter method serves to increase the particulate loading to the scrubber and is usually not cost- or energy-effective.

Scrubber types which require a sump as part of their design (Doyle or impactor types) suffer when high char loadings are experienced. Efforts to suppress the char on the liquid surface are better accomplished by introducing turbulence under the liquid surfaces than by spraying from above. Overhead spraying sometimes entrains gases which actually help the char float.

Materials of construction depend upon the fuel (wood), the auxiliary fuel (fossil), the water source (chlorinated?), and the amount of recycle. Generally speaking, mild steel construction is adequate for bark boilers with gas auxiliary when the water is from a deep well. If process water is used, a detailed water analysis must be obtained. Many times, standard 316L stainless steel is insufficient when high recycle rates increase chloride residuals to above 500 ppm. A scrubber in the Pacific Northwest is made of Hastelloy C-276 and Atlac fiberglass resin construction to resist the effects of chlorides and other halogens from wood furnish which had been floated in brackish water. The applications engineer must also consider the abrasive effects of the sand. Headers which protrude into the high-velocity gas stream will not do so for long. Either do not use them, or make them easily replaceable. Liquid inlet headers should have rod-out connections for clearing char if it should be accidently picked up in the liquid circuit.

Weirs are the best level control on bark boiler scrubbers. Electronic devices are adversely affected by char (which can produce a false reading) and sand (which can coat and foul the sensing devices). Weirs should be visible if possible, and flush headers installed if they are not. Weirs can provide a clear indication of flow and level for the operator.

Bark-, woodwaste-, or combination-fired, boiler emissions can be effectively controlled with a properly applied wet scrubber. Table 3.3 presents an application guide for bark boiler scrubbers.

3.4 FOUNDRY CUPOLA SCRUBBERS (FERROUS METALS)

The production of iron for a variety of metallurgical operations involves the use of foundry cupolas. These devices are containers for the application of heat, typically from the burning of coke, to melt previously cast ferrous pieces. Items such as engine blocks, rejects, sprues and risers, and other pieces are heated in a batch fashion (the batch charge being called the "burden") using ignited coke and limestone and a continuous blast of combustion air directed from a pressure blower through air ports in the cupola called "tuyeres." This air burns the fuel along with carbon, silicon and manganese that may be present in the metal. The melted metal is drawn off an opening called the tap hole.

TABLE 3.3. Bark Boiler Scrubber Application Guide.

Type of Scrubber: Venturi scrubber with vane separator.

Throat: Adjustable annular or rectangular; 25–100% design volume range.

Actuation: Powered manual or automatic, using breeching draft for control signal.

Materials: 316L stainless steel. Use 317L or similar alloy if chlorides are present. Suggest test coupon before composing specification.

Access Ports: Venturi approach, rod-outs on headers, flooded elbow, separator base, separator vane level and sump.

Pressure Drop: Typically 6–20 in., depending on auxiliary fuel (dry-bark gas-fired is lowest; wet-bark oil-fired is highest).

Outlet Grain Loading: 0.02 gr/scf corrected at 12% CO_2.

Inlet Temperature: Without economizer, 600–800°F; with economizer, 325–500°F.

Outlet Temperature: Wet-bark firing, 156–165°F; dry-bark firing, 150–158°F.

L/G: Recycle, 6–8 gal/1000 ft³; once-through, 4–6 gal/1000 ft³.

Droplet Control: Cyclonic or vane separator.

Inlet Velocity: Southern or coastal mills (high sand content), 50–55 ft/sec; hardwood northern mills (low sand content) 60 ft/sec.

Outlet Velocity: 30 ft/sec to stack; 60–65 ft/sec to fan.

Test Method: Front end of Method 5.

Special Consideration: High impinger catch indicates high sand content of fuel or incomplete combustion. Test when boiler is stabilized.

The combustion air is sometimes preheated so as to improve the thermal efficiency of the system. The preheating of the air tends to improve the efficiency of oxidation, causing increased emissions of fine (usually silicon dioxide) particulate.

Given the high carbon content of the charge (contributed by the coke, and by carbon in the burden and in the oil which may coat the charge), large quantities of CO are produced, along with coke-breeze, fine silicon dioxide particles (less than 0.5 μ), other metallic oxides and free carbon.

A typical particle distribution on a hot blast cupola may be:

Size, μ	% Less Than
100+	95%
50–100	90%
10–50	75%
1–10	50–65%
<1	15–30%
Specific gravity = 2.5–3.5	

This high quantity of submicron particulate results in stack tests that may have 50% of their weight catches in the impingers, even after scrubbing. The impinger catch is typically condensed hydrocarbons evolved through the incomplete combustion of the oil emitted from the burden, and silicon dioxide (from the sand reintroduced along with the sprues and risers, and also that produced from silicon in the parent metal charge).

The wet air pollution control system must:

(1) Provide for the safe ignition of the CO to CO_2 and adequate residence time for the combustion of the hydrocarbons to CO_2 and water vapor

(2) Quench the gas stream (typically near 1200°F) quickly to permit particle growth among the small particles present

(3) Avoid bends or other changes in direction which may cause abrasive wear

(4) Avoid or accommodate the effects of the use of fluxes for slag formation which could produce corrosive conditions in the scrubber

(5) Provide adjustable pressure drop control to compensate for volume changes during the melt cycle

(6) Provide adequate draft at the cupola charge door for worker safety if the cupola is hand charged or if workers are in the area

(7) Remove the particulate to opacity levels of less than Ringleman 1 (20%) or an outlet loading of 0.1 gr/dscf

(8) Neutralize the normally acidic scrubbing liquid stream (Note: baghouse or precipitator systems on cupolas do not remove the CO_2 or SO_2 fractions of the exhaust; thus, they create acidic conditions once these gases humidify.)

(9) Provide reasonable horsepower requirements in performing these tasks

Many foundary cupola wet scrubbing systems utilize:

(1) One or more after burners located in the upper section of the cupola (above the charge door) to ignite the CO, burn it to CO_2 and also burn the residual hydrocarbons. These are modulating burners controlled by a breeching thermocouple set to maintain the temperature at 1200°F or more while the blast air is on or the melt is in progress. The burners are shut off after the melt is complete and the bottom is ready to be dropped. At least 1.0 sec residence time is needed.

(2) A prequencher (spray type with wetted walls) typically mounted as near to the cupola as possible, to avoid refractory duct costs. This quenches the gases and humidifies them for particle growth and subsequent collection. It also reduces the suspended load on the building structure.

(3) An adjustable venturi scrubber operating at 25–40 in. w.c. with power positioner tied to a draft controller on the breeching. This provides

throat adjustments for draft control only in the range of venturi drops that will meet compliance. A defeat is built into the system to allow fan startup and shutdown with the throat closed.

(4) An automatic counterbalanced stack cap on the cupola to vent to atmosphere under loss of power or loss of the fan.

(5) A cyclonic separator with removable wear plates on the tangential inlet zone or a radial entry spin vane separator with at least one diameter of height from the centerline of the inlet to the face of the spin vane. A sump is usually used in the bottom of the separator to provide recycle. Overflows and drains of 2 ft/sec design are commonly used.

(6) A liquid treatment system using either a settling tank or a rubber-lined liquid cyclone (or both) for separation of the accumulated solids. A dump ("slop") bucket is usually used with a drag-chain clarifier to remove solids.

(7) A pH adjustment system using caustic and a soap breaker (the soap is formed when the caustic reacts with any residual oil) pressure-fed into the recycle pump inlet. pH probes are mounted in a tank for easy access for cleaning.

(8) Protective systems with pressure switches to monitor loss of recycle, quencher or clarifier flow. High-temperature alarms (mounted in the scrubber outlet) are sometimes used also.

(9) Vent and drain systems. Since the air pollution control device is attached to a batch operation, the scrubber must be drainable without capturing air. Many enclosed liquids/solids separator tanks must have vents installed to permit batch draining.

(10) Reliable throat actuators. The adjustable throats are typically damper blade designs with external hydraulic or air-operated actuators. This mechanism must be rugged to accommodate the high damper blade pressures which exist at these high pressure drops.

(11) Removable wear parts. The damper blades must be easily removed. Methods include an end fitting at the throat which must be removed and the blade(s) pulled out, or separate doors behind or below each blade assembly.

(12) A wet fan. The fans are radially bladed designs, direct connected with fan inlet sprays for cleaning (coupled to a timer which is defeated at startup), piped and sealed drain, and nonmetallic expansion joints at the inlet and outlet. One must remember to allow at least two diameters of straight duct run into the fan inlet for proper performance. Velocity pressure recovery evasés (fan outlet divergers) are also used on the fan outlet to slow the outlet velocity to that near the stack (30–40 ft/sec).

(13) A stack with sampling platform for U.S. EPA Method 5 testing (re-vised listing in *Federal Register,* Vol. 42, No. 160, Title 40, Ch. 1, part 60, 1977).

3.4.1 Materials of Construction

Quenchers are 316 stainless steel or refractory-lined mild steel with plug-resistant nozzles (stellite-tipped) mounted into retractable headers.

Venturis are 316 stainless steel, with some designs using silicon carbide brick on the throat damper blades. Throat actuators are hydraulic cylinder, Jordan electric actuators, Duff-Norton jackscrews or the like. Pins are recommended to stop the damper blade before it can strike the housing, if the positioner is lost during operation.

Separators are 316 stainless steel with recycle sumps designed for 10 min or more holdup.

Stacks are mild steel with a stainless steel liner or all stainless steel. Some fiberglass flake–lined stacks have been used sucessfully.

Pumps are stainless steel or Nihard. Some vendors have used rubber-lined centrifugals such as Denver or Allis-Chalmers with success.

Valves and right angle bends are to be avoided, given the high abrasive wear present.

An application guide for ferrous metals cupolas is presented in Table 3.4.

3.5 LIME KILN SCRUBBERS

Many members of the chemical processing and pulp and paper industry have operated impingement kiln scrubbers for years. Peabody, for example, has provided over 100 lime kiln scrubbers worldwide, and some are still operating after more than 20 years. In addition, venturi scrubbers are often used. Figure 3.9 shows AirPol venturi scrubbers. The Figure 3.9a cutaway shows a damper blade venturi throat in full open position. Note the tangential liquid inlets. In Figure 3.9b the high-level liquid header for minimizing solids pluggage can be seen as viewed from the feed end of the kiln. A gas inlet expansion joint appears above the venturi.

The material that follows covers first the conversion of impingement lime kiln scrubbers and then the replacement venturi scrubbers. Finally, the scrubbing system and the chemistry involved are discussed.

3.5.1 Modernizing Impingement Scrubbers

The impingement scrubbers were designed for an era of less stringent air cleaning requirements. However, they still can be used effectively today.

TABLE 3.4. Ferrous Metals Cupola Application Guide.

Typical System: Quencher followed by adjustable venturi and cyclonic separator.

Materials: 316L stainless steel; wear plates at venturi throat; replaceable venturi parts.

Piping: Avoid 90° turns; use two 45° elbows; install pipe sections in removable units. If a drag chain tank is used, allow only 35° of rake angle. Sludge tends to flow, especially if caustic is added. If a cyclone is used, 15–20 wt % slurry will result. This requires slop bucket discharge.

Fan: A fully split housing is needed for wheel removal. Vibration eliminators and detectors are suggested. The fan should be on the clean side. A sealed drain and an access door are required. Fan spray is suggested with manual operation, not with a timer. The wheel has wear plates, and scroll and check wear plates are suggested.

Pressure Drop: Spray quencher, 0.5 in. w.c.; wet wall quencher, 0.5 in. w.c.; educator-type quencher, 0 in. w.c.; venturi, 30–45 in. w.c.; separator, 3 in. w.c.

Inlet Temperature: Up to 1200°F if afterburner is used.

Outlet Temperature: Typically 156–165°F. With recycle, water will reach this temperature. Insulate worker-area pipes.

L/G: Typically 10–12 gal/1000 acfm with no caustic addition, or 12–14 with caustic addition. Add caustic to separator, not to venturi inlet. pH control is needed.

Separator Vertical Velocity: No more than 11 ft/sec.
Inlet Velocity: 50 ft/sec.
Outlet Velocity: 30 ft/sec with stack; 60 ft/sec to fan.

Test Method: Method 5, with the front half only counted.

Special Considerations: Flooded elbow after venturi. Caustic addition plus foam suppressant for oily scrap problems. Liquid circuit uses settling tank or liquid cyclone (rubber-lined or Nihard) for solids rejection. Extra-large manhole (24 in.) is needed in flooded elbow for maintenance. Cyclonic separator uses wear plate for first 270° of separator shell at inlet. High back half indicates sand on returns and silica emission. Cleaner scrap can reduce emissions. Oily emissions will cause a blue-haze residual plume. Afterburners can help remove the haze, but a scrubber will not.

Their clever use can save thousands of dollars in replacement and installation costs. One can recycle a higher percentage of solids, reduce clean-out periods and improve stack emissions, and it is relatively simple to do.

Most impingement lime kiln scrubbers fortunately were designed with a vertical velocity within the range of cyclonic separators; that is to say that an impingement scrubber, with all internals removed, can be converted to a full cyclonic device. Given this good fortune, a modern annular venturi can be added to the unit, giving it up-to-date removal capabilities. Of course, the fan energy required will increase, but only to a level equal to or less than similar designs (see Figures 3.10 and 3.11).

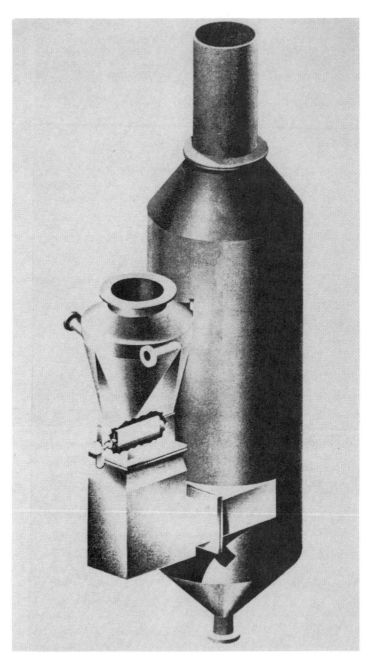

FIGURE 3.9a. Lime kiln venturi gas scrubber (courtesy Neptune AirPol, Inc.).

106

FIGURE 3.9b. Lime kiln venturi gas scrubber (courtesy Neptune AirPol, Inc.).

The steps to be taken involve:

(1) Inspection of the existing scrubber
(2) Measurement
(3) Sizing of the scrubber
(4) Fan selection
(5) Installation

3.5.1.1 INSPECTION OF THE SCRUBBER

The existing scrubber is inspected for material corrosion and mechanical integrity. The manholes are removed and inspected as are any easily accessible internals. The thicknesses are tabulated and compared to original Peabody details of construction.

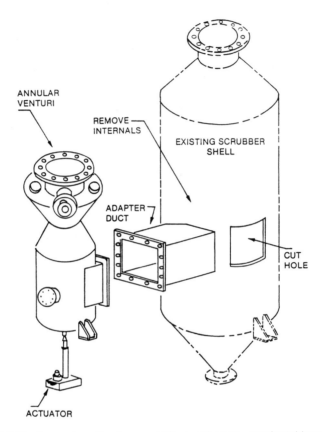

FIGURE 3.10. Conversion of impingement kiln scrubbers to venturi scrubbers (courtesy Neptune AirPol, Inc.).

If the scrubber will be attached to an oil-fired kiln, the unit should be stainless steel or of a diameter sufficiently large for nonmetallic lining, such as precrete or gunnite. If a gas-fired kiln is used, mild steel will suffice.

Tray and tray supports should be inspected thoroughly for means of attachment. Almost all trays are removable, but the tray supports are welded in place.

The external recycle tank, if used in the original design, should be sized thoroughly to check its hold-up capacity. Even this tank will remain.

The pump nameplate data should be recorded. The new scrubber may use slightly more recycled liquid but will give the added benefit of the ability to recycle a higher solids concentration slurry. The new annular will tolerate 10–15 wt % of solids.

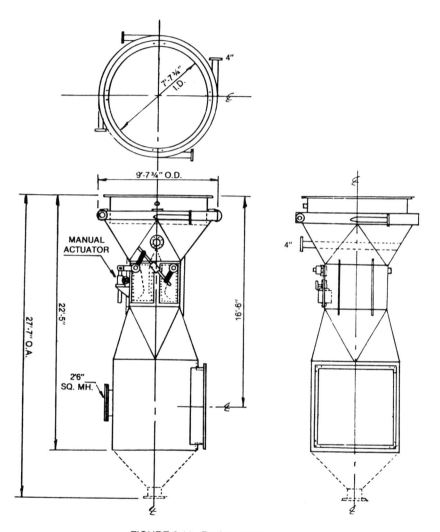

FIGURE 3.11. Replacement venturi.

3.5.1.2 MEASUREMENT

The internal diameter of the vessel is measured, as is the height from the bottom tangent line (where the conical sump meets the side wall) to the centerline of the gas outlet (or the top of the gas outlet flange if a top central outlet device). The area for the venturi is also checked.

3.5.1.3 SIZING OF THE SCRUBBER

Engineers then compare the data generated with the original project drawings and the new dimensional requirements. Features of an annular venturi are shown in Figure 3.12.

The venturi fits in the inlet duct to the scrubber by way of a special

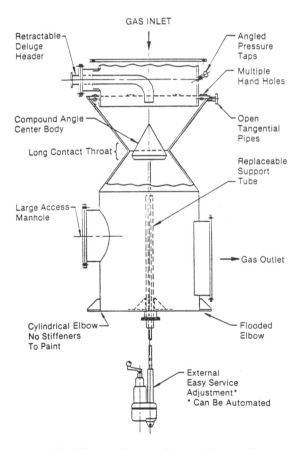

FIGURE 3.12. Features of an annular venturi.

adapter duct. This adapter duct is to be welded tangentially into the old vessel. The adapter duct provides a carefully calibrated tangential inlet to the old separator, thus providing the centrifugal force which separates the liquid droplets from the scrubbed gas stream. Since this adapter is made as a unit, it can be easily replaced as a unit over the years of service the system will provide.

A neoprene gasket is provided to seal the inlet adapter to the venturi. The bolts, nuts and washers are also provided so that field installation is minimized.

3.5.1.4 FAN SELECTION

The new annular venturi uses more energy simply because it takes more energy to remove more particulate. The annular venturi responds more uniformly to positioner location than a damper blade design or similar pivoted arrangements. When a damper is pivoted, the area it exposes to a gas stream varies geometrically with the air stream. (Since one end is fixed, it presents its surface at an angle to the air flow.) Thus, when the damper blade venturi approaches its most fully closed position, it has the least control. The slightest movement of the blade will cause pressure drop changes, sometimes enough to upset the kiln. It also wastes energy.

The annular venturi presents a constant frontal area to the air stream. Its pressure drop increases more uniformly, creating exceptional control for the operator. This is why many successful high-energy (60–80 in. w.c.) venturis are annulars. The pressure drop across the venturi approaches the minimum possible for a given removal efficiency.

Given the code requirements, lime mud soda content, length of kiln, whether the kiln is burning noncondensibles, and type of kiln fuel, the engineer can select the proper pressure drop for the given application. The kiln draft loss and the feed end housing loss are added to the venturi drop plus 3 in. w.c. separator loss, giving the total fan static required at operating conditions. This amount is typically 26 to 32 in. w.c.

If a *hot fan* (fan ahead of venturi) is used, then mild steel construction can be used with wear liners on at least the blade and preferably on the cheek and scroll. The fan can be direct-connected or variable-speed depending on the load fluctuation of the kiln expected during the year. Heat slingers are used and arrangement 3 fans are typically recommended in single-inlet construction up to 25,000–50,000 acfm and double-inlet above 50,000 acfm if possible.

Opposed blade inlet dampers with extended grease fittings are suggested for the fan, as are drains, access doors and shaft sleeves. Asbestos shaft seals are sometimes used.

Wet fans are used after the separator, and must be stainless steel on oil-fired kilns. They should have the same trim as the hot fan, with the exception of the heat slinger, extended grease fittings and wear liners. Since they blow cleaned air, attention to flushing should be of more concern than wear. Usually an inlet spray and housing spray are used to periodically clean the unit.

3.5.1.5 INSTALLATION

The old scrubber is first cleaned of large lime buildup and permitted to air-dry. The trays are removed during this period, as are the tray spray headers and flushing nozzles.

The tray supports and vane eliminators are then totally removed from the vessel and metal remnants are ground flush with the vessel wall. Minor patching is done at this time.

The upper rectangular manholes are totally removed and their opening replaced with plate, so that a smooth cylindrical vessel results. The lower manhole is retained for access to the sump, or, if in the wrong position, it too is removed and replaced with plate, and another manhole added.

Any internal baffles or head pipes are removed from the sump, leaving just the level control connection (if used), the makeup water connection, overflow and drain. The fan is set into place and assembled. Now the rebuilding starts.

The adapter duct is fitted into a hole cut tangentially into the side of the vessel. The existing inelt is removed and blanked off. In some cases, even this inlet can be saved, becoming the new adapter duct. It is not welded in place yet.

The new venturi is added ahead of the adapter duct, and is located on structural steel and bolted to the adapter duct using the gaskets and bolts provided. The venturi is checked to be sure that it is vertical. Shims are added to obtain a vertical centerline. Then the adapter duct is welded in place, both inside and out. The venturi is then piped in accordance with the flow diagram. The piping is typically less complex than the original unit. Modification to the pump, if any, is achieved at this time. The old pipe, usually similar in size, is reused if possible. The venturi will require duct-work modification to connect either to the feed end housing or to the fan outlet depending on fan location. The mill usually does the duct-work itself, subcontracts, or provides slip ring duct connections for field fit-up and adjustment. Expansion joints are retained as much as possible. The unit is then started up in accordance with the written startup instructions provided.

It can readily be seen that modification of an existing structurally sound Peabody scrubber can save significantly in expense and in installation time. It extends the useful life of an investment without depletion of as much valuable capital. In addition, it is a maintenance procedure which can have tax advantages over capital purchase.

3.5.2 Selecting Venturi Pressure Drop

An approximate pressure drop can be determined quickly if three important pieces of information are available. These are:

- the lime mud soda content, expressed as Na_2O in % kiln mud feed
- kiln outlet dust load, lb/min
- kiln mud feed, dry basis, lb/min or ton/day

The sodium compounds in the mud feed to a lime sludge kiln are assumed to completely burn to Na_2O in the calcining zone of the kiln. Unfortunately, Na_2O is in a fine submicron fume which represents the majority of the particulate emissions from a scrubber. The lime itself is composed of fugitive particles that are quite large and relatively easy to remove in the scrubber.

3.5.2.1 PROCEDURE

Determine lb/min of Na_2O fume.

(1) Take the total dry mud feed and multiply by the soda content of the mud. All of this weight is assumed to be fume coming out of the kiln.

(2) Look at Table 3.5 to determine the expected percent removal at a given pressure drop. Pick a "test" pressure drop as an initial trial.

Determine lb Na_2O removal.

(3) Multiply pounds of Na_2O determined by the % removal. This gives the Na_2O removed. Now subtract the amount removed from the amount of Na_2O entering. This gives the amount of Na_2O still leaving the unit. Store this number.

Determine lime removed.

(4) Take the kiln outlet dust load and subtract the amount of Na_2O. This gives the weight of lime being emitted by the kiln.

(5) Check Table 3.5 for lime removed at the same pressure drop used in step (2). Take this percent removal and multiply by the lime emission weight. This gives the lime removed and retained in the scrubber prior to bleed. Subtracting this figure from the inlet lime emission gives the amount of lime emitted up the stack after scrubbing. Add this lime emission to the soda emission and determine, from the code or customer requirement, if your selected pressure drop is adequate. If it is not, repeat the procedure. Figure 3.13 shows the venturi ΔP safe range.

TABLE 3.5. *Venturi Pressure Drop vs. Soda Fumes
and Lime Dust Removal on Paper
Industry Lime Kilns.*

Venturi Pressure Drop, in. w.c.	E_s, % Na_2O Removed	E_l, % Lime Removed
8	50	98.00
10	62	99.00
12	70	99.10
14	78	99.40
16	82	99.50
18	85	99.60
20	88	99.65
22	90	99.75
24	92	99.80
26	95	99.85
28	96	99.90
30	97	99.95

3.5.2.2 EXAMPLE

M = mud feed, lb/min dry
S = % soda, lb Na_2O/lb feed
E_s = soda removal efficiency, %
E_l = lime removal efficiency, %
K = outlet dust loading, lb/min lime and soda

Solution:

(1) $M \times S$ = lb/min of Na_2O fume coming from kiln = F

(2) $F \times E_s$ = lb/min Na_2O captured = F_c

(3) $F - F_c$ = lb/min Na_2O lost = X

(4) $K - F$ = lb/min lime dust emitted = L
    $L \times E_l$ = lb/min lime dust captured = L_c
    $L - L_c$ = lb/min lime dust lost = Y

(5) $X + Y$ = total dust emitted, lb/min = Z
    $F_c + L_c$ = lb/min lim + Na_2O captured in scrubbing liquid

If Z exceeds code, a new, higher pressure drop is tried. Typical kiln pressure drops are 26–28 in. w.c. to meet new source standards and 18–20 in. w.c. on retrofit installations.

A lime kiln scrubber application guide is presented as Table 3.6.

3.6 MUNICIPAL SLUDGE INCINERATOR SCRUBBERS

The control of emissions from multiple-hearth furnaces involves the application of fine particulate abatement equipment, usually followed by gas cooling. This results from the fact that the sludge is fed into the furnace at high moisture content (sometimes as high as 40%), yielding a high moisture content saturation temperature plume. Saturation temperatures in the range of 165–170°F occur, producing dense high-moisture (and very visible) plumes. Figure 3.14 shows a composite particle size distribution.

Typical techniques involve the use of a medium-energy (20–28 in. w.c.) venturi scrubber followed by a tray or packed aftercooler. Since the sludge moisture content can vary significantly, the flue gas volume is subjected to

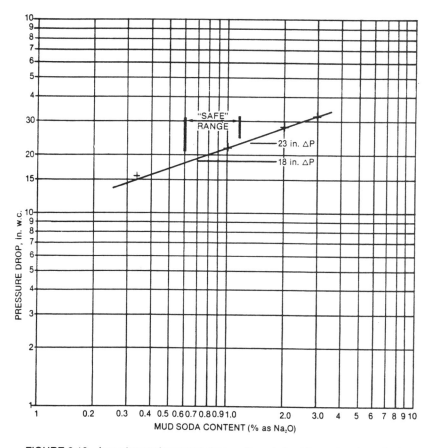

FIGURE 3.13. Annual venturi pressure drop vs. lime sludge kiln mud soda content.

TABLE 3.6. Lime Kiln Scrubber Application Guide.

Type of Scrubber: Annular venturi.
Throat: Adjustable high turndown.
Actuation: Remote manual from control room – avoid automatic designs.
Materials: 316L stainless steel.
Access Ports: Liquid shelf, flooded elbow (at least 18-in.-dia. manhole), separator, drain rod-out.
Pressure Drop: See graph in this section, typically 26–36 in w.c.
Outlet Grain Loading: 0.02 gr/scf (lower values have been reported).
L/G: 12–16 gal/1000 ft³, recycle and bleed.

Droplet Control: Cyclonic separator.
Inlet Velocity: 50–60 ft/sec.
Outlet Velocity: 60 ft/sec to fan (wet fan design); 30 ft/sec to stack.

Test Method: Method 5.

Special Considerations: Use open headers, no spray nozzles. Feed all headers from below, permit to drain into scrubber, allow clean-out tee at separator base and/or at recycle pump suction. Close throat fully for kiln startup; open gradually to avoid loss of flame. Check mud soda content, control as best as possible through good mud washing. High soda yields high emissions. Potential metallic oxide emission when burning No. 6 oil. Inspect scrubber yearly.

significant swings. It is for this reason, among others, that adjustable venturi scrubbers are used. This unit provides the operator with an adjustment on the venturi pressure drop *and* on the furnace draft. By opening the throat, increasing draft can be placed on the multiple hearth. This draft is usually regulated to the manufacturer's requirements since this helps maintain burner infiltration at the optimum level and reduces aerodynamic suspension of sludge particles (carryover). This draft may be 0.1–0.2 in. w.c. to as high as 0.7 in. w.c. if the furnace uses a knockout device. Placing a high draft on the multiple hearth is to be avoided if one expects to reduce the inlet grain loading to the scrubber.

The venturi scrubber is, many times, a wetter approach design, perhaps with a bulk spray to augment the flow of scrubbing liquid to the throat. Spray throat venturis are to be avoided, since the efficiency can drop off significantly in the event of nozzle plugging or loss of header flow. The throat itself should be rectangular, or annular if the volume is over 60,000 acfm. If rectangular, the throat effective zone must be no wider than 9 to 12 in. If it is wider, the liquid cascading down the throat approach will not progress to the throat center, creating an ineffective void.

Simple throats are best for this application. The throat should be flanged for future replacement since it is a wear zone. Designs which permit the replacement to the throat damper blade alone are fine; however, typically the entire throat needs replacement when the blade shows deterioration. The throat should be manually articulated, with remote actuator if the installation location does not permit access. One must keep in mind that throat velocities of 200–300 ft/sec are present in this type of scrubber, so it is wise to counterbalance the throat adjustment to facilitate movement. Unbalanced throat shafts may require high torque to move the throat damper.

Many designers feel the use of a bulk spray nozzle (a ramp bottom full- or hollow-core nozzle) aimed at the throat zone from above helps make certain that droplets are properly produced in the throat. This header should be bayonet-mounted so that it may be removed, and should incorporate a pressure gauge on the header to denote nozzle plugging. If a bulk nozzle is used, the wall water can be reduced to 2 gal/1000 acfm if the approach is a 50–60° hopper angle. This quantity must be administered so that the approach is uniformly covered.

The venturi throat should receive about 7–12 gal of recirculated scrubbing liquid per 1000 acfm. Solids content, by weight, should not exceed 2–4%, not because of the plugging potential, but because the inlet gas can spray dry some of the solids out of the scrubbing liquid, effectively raising the grain loading to levels higher than those measured at the multiple-hearth outlet.

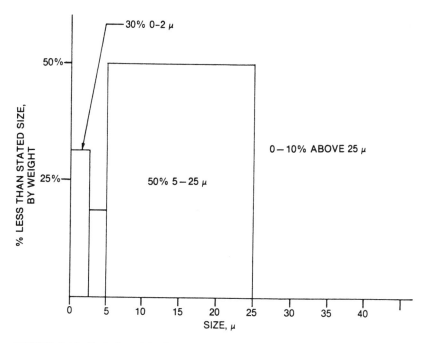

FIGURE 3.14. Typical particle size distribution for a sludge multiple-hearth incinerator (courtesy Schifftner and Associates).

A compound-angle elbow or a flooded elbow can be used. The particulate is not excessively erosive, but it can be corrosive. Organic acids can form, and will condense upon application of water vapor and transfer of sensible heat to both the surroundings and the scrubbing liquor. Many applications require 316L stainless steel construction unless the water applied to the scrubber is chlorinated. The designer and user must pay the added cost of using 2.75% moly or Inconel instead of 316L stainless steel if high chloride loadings exist. The typical material used is stainless steel, however.

A manhole of at least 18 in. diameter should be used in the elbow and handholes provided near the liquid inlets. Rod-out plugs should be used near nozzles if used, and the headers should feature flanged construction or unions to facilitate maintenance. These header assemblies should be small enough for one man to manipulate, if possible.

The separator can be cyclonic; however, if a packed or tray cooler is used, the unit can be radial. If the separator is cyclonic, at least 1-1/2 diameters of vertical shell rise should be measured from the centerline of the separator inlet to the absorber/cooler. This provides adequate disengagement. Separator tangential velocites of 80 ft/sec are common if followed by a cooler.

Spray-type aftercoolers are sometimes utilized. These modified cyclonic scrubbers spray water tangentially in a uniform pattern, providing high-surface area droplets for heat transfer. Obviously, they rely upon properly functioning nozzles, adequate pressure and flow rate. Loss of any function will cause a dramatic drop in efficiency and a noticeable rise in outlet temperature.

Tray aftercoolers are designed to drop the stack temperature from about 170°F to 115–130°F depending on the application. Table 3.7 shows how much water vapor 1 ft³ of saturated air contains at 170°F vs. what it contains at 115°F. If a fan is used on the wet side (after the scrubber), the unit can be smaller since the airflow is reduced.

Tray aftercoolers use a group of flooded impingement plates to provide the liquid surface area required for cooling. The number of plates (trays) required is calculated as in typical gas cooling calculations. Actual trays are about 75–85% of a theoretical plate. Therefore, the number of impingement

TABLE 3.7. Water Vapor Content of Saturated Air.

Temperature	Weight of Water Vapor, lb/ft³ Saturated Mix	Density (lb air + water vapor/ft³)
115°	0.004315	0.066
170°	0.01612	0.05343

trays required is found by dividing the number of theoretical plates (HETP) by the fractional efficiency, ϵ:

$$N_T = \frac{(\text{HETP})}{\epsilon}$$

Most designers allow for higher-than-normal liquid temperatures, usually 50–70°F inlet, with a liquid discharge no closer than 10° to the final gas outlet temperatures. Thus, the water is assumed to be less than 100% efficient (where it would leave at exactly the gas outlet temperature). This amounts to a maximum efficiency of approximately 90%, even when the inefficiency of the tray itself is adjusted.

The trays can be installed in stages in single-flow arrangement (liquid enters one side, flows across, descends through a weir seal, and on to the next tray), split-flow (liquid enters a center weir, flows both right and left, descends through weirs to the next tray), or variations of the two.

Superficial or face velocities (i.e., calculated on empty chamber cross-sectional area) are usually 500–600 ft/min maximum. The tray unit is sometimes followed by a restriction in the shell to accommodate a vane-type droplet eliminator. This design has open velocities of 1200–1900 ft/min and uses vanes to impart a centrifugal force to the air stream. The higher-density water droplets are propelled against the vessel wall, impact, and drain back.

An improved method utilizes a chevron droplet eliminator, sprayed intermittently from below for cleaning purposes. This unit must have support beams and manholes large enough to install the chevron in sections.

The packed tower cooler usually requires the equivalent of 4 ft of 2-in. packing (Intalox Saddles) per actual impingement tray. Since the tray-type unit only requires 18 in. per tray, a 30-in. height penalty is assessed against the packed cooler for each stage. Typical packed coolers require 6 to 8 ft of packing with a spray-type distribution. A support grid is used, and frequently a redistributor is placed in the packing to ensure that the liquid is properly mixing in the packing.

Packed coolers require manholes at the top and bottom of the packing, with the lower one typically larger (24 vs. 18 in.). The only method of cleaning the packing involves removal of the packing.

Tray scrubbers use about 2 gpm per ft of weir, which yields liquid-to-gas ratios of 3 to 4 gal/1000 acfm. Packed coolers use more water (for distribution purposes) and are on the order of 6 to 8 gal/1000 acfm.

The result is a predominance of tray-type aftercoolers; since the vessel is smaller, the maintenance is simplified and the liquid requirements are closer to theoretical.

Table 3.8 is an application guide for municipal sludge incinerators.

TABLE 3.8. Municipal Sludge Incinerator Application Guide.

Type of Scrubber: Venturi.
Throat: Annular or rectangular adjustable.
Actuation: Manual or draft automatic.
Materials: 316L stainless steel.
Access Ports: Flooded elbow manhole.
Pressure Drop: 25–36 in. w.c.
Outlet Grain Loading: Less than 0.08 gr/scf at 12% CO_2.
Inlet Temperature: 800–1400°F.
Outlet Temperature: 100–125°F with aftercooler.
L/G: 8–12 gal/scf.

Droplet Control: Separator.
Separator Type: Vane or cyclonic.
Inlet Velocity: 60 ft/sec for vane; 90 ft/sec for cyclonic.
Outlet Velocity: 30 ft/sec for stack; 55 ft/sec for fan.

Test Method: Method 5. ORSAT test for CO_2.

Special Considerations: Impingement tray cooler used to suppress plume. Stainless steel trays. Draft control override required for startup. Codes typically production-dependent measured on dry solids basis, difficult to meet if high in dissolved solids or moisture.

3.7 SEWAGE SLUDGE DRYER SCRUBBERS

Long before the recent "energy crisis," the challenging problem of sewage sludge disposal confronted the engineering community. When energy prices began their sudden rise, the tentative solutions to waste disposal had to be viewed in a new light. The questions changed from "How do we dispose of wastes?" to "How can we attain efficient low-energy waste disposal?" The answer to the first question appeared to be landfill in its variety of forms and/or incineration. The answer to the second is yet to be found. Much emphasis, however, is now being placed on *drying* for reuse or subsequent incineration. The arguments for drying of sludge include, but are not limited to the following:

(1) Drying has lower energy requirements, since lower temperatures are used in drying than incineration. Water is not "incinerated" with the sludge, it is driven off as low-temperature water vapor and steam.

(2) Varying flows can be accommodated.

(3) The end product may be useful as a fertilizer, for anaerobic digestion to produce methane, or a stored dried fuel (bulk or pelletized) for subsequent incineration.

(4) Lower scrubbing energies are needed. Low-temperature drying produces larger, more easily removed particulate.

We will deal with the air pollution control aspect of this problem—what the scrubber design should incorporate, how it should function, what the energy requirements are and what performance would be expected. The process will be observed from the scrubber design point of view.

3.7.1 Sludge Drying System

The sludge drying process involves these basic components:

- dewatering press or filter
- wet storage bins
- return product bins and mixer
- rotary drum dryer
- cyclone product separator
- induced-draft fan
- wet scrubber
- odor control system

Figure 3.15 is a typical flow diagram.

The wet pressing of the sludge has two advantageous effects on the scrubber. First, the water vapor carryover in the dryer is lower, making the scrubber smaller than it would be if the flow were incinerated. Second, the water discharged by the press would carry away some dissolved solids that would ordinarily be a pollution emission unless removed. By efficiently pressing the sludge, the scrubber can be smaller and more efficient, using less energy per pound of dried product.

In the rotary dryer, hot gases from a furnace evaporate the water from the tumbling sewage sludge product. Unexposed to the furnace gas stream, the chances for the sludge to form odorous, partially burned hydrocarbons are reduced considerably. Careful dryer design in the furnace/drum area reduces localized hot spots and partially eliminates burning of product. Major dryer manufacturers such as Zimpro, Stearns Rogers, the Heil Company and others have developed the design parameters to a point where uniform drying can be reliably attained.

Instead of exit temperatures of 350–500°F as in an incinerator, the exit temperature from the dryer is typically 180–250°F, with common operation in the 180–210°F range. By keeping the temperature low, the higher boiling point compounds in the sludge are not vaporized, thus reducing the odor loading to the scrubber system. The lower temperature also reduces the types of odors which would be emitted, i.e., amines instead of amine oxides, hydrocarbons instead of aldehydes, and nitrates instead of NO_x.

This is a key point. In an incinerator, the odors are removed by thermal oxidation at extended elevated temperatures. Partial or insufficient combustion can produce odorous compounds that may create periodic or con-

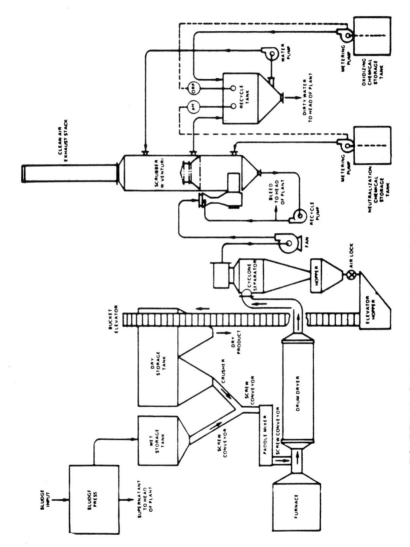

FIGURE 3.15. Sludge dryer system (courtesy Peabody Engineering Corporation).

tinuous problems. In a dryer, however, the odors are evolved at lower temperatures, and chemical oxidation rather than thermal must be used to remove them. The most common resulting odors emitted from the dryer are:

- H_2S
- amines and ammonia
- organic acids (volatile)
- indole and skatole
- trace mercaptans and sulfides
- disulfides and trace aldehydes

Most of these compounds are removed with either neutralization or oxidation or a carefully designed combination of the two. We will discuss that later regarding odor control.

3.7.2 Dryer Particulates

The dryer exhaust is sent to a high-efficiency long-tube cyclone collector or group of collectors depending on the exhaust volume. With most dryers, the particulate is large, $100 + \mu$, and has a high grain loading (18–30 gr/acf) which lends itself well to cyclonic removal. Unfortunately, the particulate is considerably friable, that is to say, it breaks down to smaller particles through impaction or abrasion on surfaces. Thus, the cyclones have to have fairly low inlet velocities and low body velocities. Inlet velocities of 55–65 ft/sec have been used on some dryer cyclones with success.

Design features of the cyclones for long life and simplified maintenance should include, among other things:

- abrasion-resistant replaceable liner
- total flanged cyclone top section to facilitate access
- minimum of 90° involute entry for gradual turning to reduce breakup
- long disengaging tube
- baffled hopper to prevent short circuiting in multiple-cyclone arrangements
- flanged body sections with lugs to remove sections more simply

Gaskets can be full-faced neoprene or asbestos with stainless steel bolts and washers. The cyclone body should be stainless steel; however, insulated carbon steel has been used on similar applications. Since the dryer emits nearly saturated air with organic acids, care should be taken to insulate all nonstainless pieces of equipment to reduce chances of condensation. Dryer installation instructions regarding insulation and access to ductwork sections should be strictly adhered to.

3.7.3 Scrubbers

Since the particulate is large going into the cycle, extremely high cyclone efficiencies are reported (e.g., in the 90+ % range). With an average inlet grain loading of 20 gr/acf, this leaves only 0.4 gr/acf entering the scrubber. Scrubber designers are encouraged to use 0.5–1.5 gr/acf in their design since fluctuations can easily cause increased grain loading.

The goal is to remove as much particulate as possible in an effort not so much to meet pollution codes (this is relatively easy), but to reduce chemical demand in the subsequent odor control stage. Every pound of dust not captured and removed in the venturi will require some chemical to neutralize or oxidize it in the odor control stage. The efficiency curve given in Figure 3.16 shows the approximate outlet loading vs. pressure drop. At only 10–14 in. w.c. with a properly designed scrubber, nearly complete particulate removal can be attained on a 24 hr/day basis.

The venturi scrubber most used on this application features an adjustable throat with nozzle-free design such as that shown in Figure 3.17. That is to say, no spray nozzles which may plug are relied upon to produce any scrubbing effect. The scrubbing liquid is aerodynamically oxidized in the high-velocity venturi throat as a result of wide differences between the velocity of the air and that of the scrubbing liquid. This shearing action creates small droplets much like the spray produced by strong ocean winds blowing across wave tops.

The lack of nozzle plugging problems could indicate that the nozzles are worn out. This results in inadequate gas-liquid contact which should concurrently produce a drop in efficiency. Since the odor control stage which follows is automatically controlled, any nozzle problems will cause a subsequent (immediate) increase in chemical, adversely affecting the economics of the system.

A round, fully wetted venturi approach or similar design is recommended, as is an adjustable venturi throat. By being adjustable, the dryer draft can be controlled at two places – the venturi and the fan damper – thus providing high scrubbing efficiencies and uniform draft. When the venturi has reduced draft losses, the increased draft will cause excessive product carryover, burner problems in the dryer furnace, chemical addition increases and higher solids removal rates from the scrubber. It is clear to see why an adjustable throat is recommended.

3.7.4 Mist Elimination

Existing systems vary greatly in the method by which liquid droplets from the scrubber are separated from the air. The intent is the same: to remove as many droplets (which contain the dust particles) as possible to

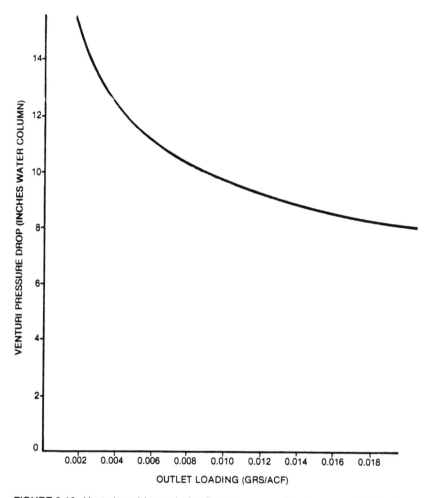

FIGURE 3.16. Venturi scrubber outlet loading vs. pressure drop for sewage sludge dryer scrubbers.

prevent carryover into the odor control stage. One design uses chevron baffles and the other cyclonic action. Cyclonic action, in the opinion of many, is preferred since (1) less surface area is provided upon which organic material may build, thus reducing chances of odorous decomposition after a period of operation, (2) there are no internals to replace or clean, and (3) it costs less. Chevron baffles would be better used on higher-energy scrubbers where the droplets are smaller. Low-energy scrubbers produce large droplets which lend themselves to cyclonic separation. Only about 1 in. w.c. additional energy is required. After the separator, particulate loadings

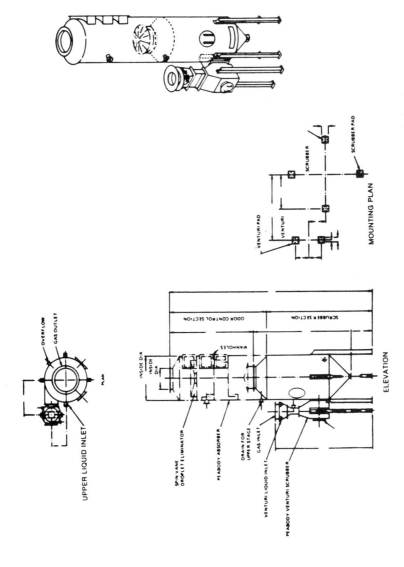

FIGURE 3.17. Adjustable-throat venturi scrubber with absorber for sewage sludge dryer.

126

of a mere 0.01 gr/acf or less are common. The liquid and solids removed are sent back to the treatment plant headwaters, or other convenient locations.

3.7.5 Odor Control Stage

Now that the particulate has been removed, we must deal with the gaseous pollutants, some of which are odorous. After passing through the liquid/gas separator, the gases pass through a baffle which reduces liquid carryover from the venturi scrubber to the absorber. This baffle can be in a variety of shapes; however, it must be free-draining and of open design to reduce back pressure. Annular velocities of 20 ft/sec are common.

The gas then passes to the absorber. Since the chemical reactions occur in the liquid phase, the gas must be absorbed first so that maximum odor removal can occur. Gas phase reactions such as Cl_2 addition are not typically considered because of increased corrosion possibilities and the increased likelihood of forming toxic chloramines. ClO_2 addition is a possibility, yet under high concentrations it reacts violently with ammonia. It is effective, however, in the liquid phase and may be advisable where a high percentage of industrial-based sludge is dried. Chemicals of interest are hydrogen peroxide with an iron salt catalyst, or potassium permanganate in a buffered solution.

Hydrogen peroxide decomposes to H_2O and O_2 and has been proven for wet well odor control and injection into force and gravity mains. Key factors here are reaction times. Hydrogen peroxide requires 3–5 min of retention time to thoroughly oxidize H_2S. When used in a slightly acidic solution (pH 6.5–7) it will produce elemental sulfur on a pound for pound basis. When using an acidic scrubbing liquid, the ammonia present in the gas stream will be absorbed and neutralized. Unfortunately, domestic sewage typically tends to raise the pH of water when dissolved. Therefore, the first stage scrubber (venturi) should have pH control and acid addition to remove soluble ammonia. Rather than use HCl, which would introduce chloride ions into the circuit (H_2O_2 dechlorinates water), commercial-grade sulfuric acid could be better applied.

It is recommended that the pH control be placed in the absorber circuit so that a rise in pH will indicate reduced ammonia carryover, thus triggering increased acid dosage to the pump inlet of the first stage. This places strong acid where it will be most effective, in the venturi, while monitoring it at the place of greatest effect, the absorber. An oxidation-reduction potential (ORP) meter in the upper stage adds H_2O_2 and catalyst only on demand, thus conserving chemical. If permanganate were to be used, the pH would have to be varied by experimentation with a "first choice" of 8.5–9. Unfortunately, permanganate leaves a manganese dioxide residual which can present operational problems in packed absorbers or trays. Permanganate is considerably more expensive than many alternative oxidants.

Since most of the odorous emissions are hydrophilic, a standard absorber is recommended. Packed beds have been used in the past on manure dryers; however, tray absorbers like those used on sludge incinerators and on a variety of dryer designs offer greater height and liquid flow savings. They have the following advantages:

- Only 3–5 psig liquid pressure are needed.
- There are no spray nozzles to plug.
- They have been used extensively on sludge incinerator scrubbers with success.
- Trays are easier to remove and maintain than packing.
- They require less vessel height.
- They offer very high absorption rates and low flow rates.
- They are readily available.

Packed bed depths of 4–6 ft of 2- to 3-in. Intalox saddles or two impingement trays are sufficient. Liquid rates of 12 gal/1000 acfm for the packed beds and 4 gal/100 acfm are common for each respective device.

The gases then pass through a mist eliminator. Since no spray nozzles are used in a tray absorber, chevron baffles are sufficient for droplet removal prior to discharge. In the case of a spray-distributed packed tower, a mesh-type mist eliminator may be advisable, but the design must allow for periodic solids backflush.

The gases are now essentially odor-free and contain less than 0.01 gr/acf of particulate. However, the emission is saturated with water vapor at approximately 150–170°F. It may be advisable in some instances to add an aftercooler consisting of one tray of 3 ft of packing after the absorber to reduce plume emissions and retain water vapor. This may be especially important in metropolitan areas where plume nuisance problems may occur.

The total system drop for a packed bed design would be 16–18 in. w.c. and 14–16 in. w.c. for a tray design, measured from inlet flange to outlet flange. An incinerator scrubber would be 30–36 in. for a similar volume.

The drying of sludge has significant advantages over alternative waste disposal systems. The wet scrubber applied to the dryer is more efficient than its incinerator counterpart and uses less energy since the particulates are larger, the temperature lower and the odors of lower volatility. The scrubber design is extremely important in the analysis of this alternative. Table 3.9 is a sludge dryer scrubber application guide.

TABLE 3.9. Sludge Dryer Scrubber Application Guide.

Type of Scrubber: Venturi, rectangular throat.
Throat: Adjustable damper blade.
Actuation: Manual.
Materials: 316L stainless steel.
Access Ports: Liquid shelf, removable headers.
Pressure Drop: 8–14 in. w.c.
Outlet Grain Loading: Less than 0.05 gr/scf.
Inlet Temperature: 180–250°F.
Outlet Temperature: 165–178°F (high outlet humidity).
L/G: 6–8 gal/1000 acf.

Droplet Control
Separator Type: Cyclonic with second-stage absorber (packed or tray).
Inlet Velocity: 90 ft/sec tangential velocity.
Outlet Velocity: 30 ft/sec for stack; 55 ft/sec for ductwork.

Test Method: Method 5, analyze impinger catch for specific odors or residual
    materials.

Special Considerations: Abrasion—moderate; wear plate on separator inlet. Cor-
    rosion—high; pH control on both stages or water once-through.

3.8 INDUSTRIAL BOILER SCRUBBERS

The recent increases in the cost of refined oil and gas as industrial heat sources have prompted engineering reviews and implementation programs designed to utilize more readily available and/or cheaper fuels. These fuels, such as No. 6 oil or coal, cause greater particulate and gaseous pollution when burned than do refined oil and gas.

The reduction of the effects of these pollutants has been addressed through improved combustion processes and controls and by thoroughly designed cleanup equipment. The former may entail feedback controls of excess air levels, using a flue gas oxygen or combustibles detector, precise temperature controls, fuel precleaning, increased turbulence and a host of other combustion process adjustments.

The design of air pollution control cleanup equipment is much more involved. The designer is confronted with problems in three important areas:

- design of the particulate removal system
- design of the sulfur dioxide control system
- the disposal of the wastes

The design starts at the conditions for ultimate disposal of the waste—not at the heat source breeching. One must design the system so that the waste materials may be adequately disposed of for the particular user's plant.

A factory located in the middle of a densely populated area would not be able to utilize a system which requires large settling ponds for sludge separation and disposal. It would require, however, a highly effective control system given code requirements and the visibility its location provides. The designer would have to provide a compact, effective cleanup system with a cart-away waste disposal system. The engineer would have to investigate local means of disposal, its cost, its complexity, and its risk of sudden termination. He would have to produce an economic analysis of the costs to his firm for the switch to lower-cost fuels accommodating all of the "hidden" charges which may occur.

Designers of a plant located in a rural area would have to investigate the cost and complexity of importing chemicals for the SO_2 removal system, future expansion requirements, space for ponds for sludge disposal, storage areas for reagents, handling costs, and other factors which may impact on their decision regarding the type of control system best suited to the plant's operation. They may find that switching to cheaper fuel is actually more expensive. They may find that the installation of a cogeneration facility using clean-burning fuels will enable them to reduce purchased power costs, yielding a lower net cost for operation. In many instances however, the data, engineering and economic, point to conversion.

3.8.1 Particulate Removal

Industrial boiler particulate removal systems involve cyclone collectors, venturi scrubbers, cyclonic scrubbers, baghouses and precipitators. Boilers burning coal usually use knock-out chambers and/or cyclone collectors for particulate control. Large-diameter cyclones typically remove particles of 50 μ and larger and do not break these particles into smaller sizes very readily. Small-diameter multicyclones (4-, 6-, 8-, 10-, 18-in. tubes) removes particles down to 10–20 μ, but tend to break larger particles into much smaller size ranges. It is not uncommon to find the particle size distribution for a multicyclone-equipped boiler favoring particles less than 10 μ. On a mass emissions rate basis, the multicyclone unit removes more particulate; however, it does place burdens on the collector that follows. One therefore favors large-diameter coarse-particulate arrestors preceding scrubbers rather than small-diameter fine-particulate collectors. Unfortunately, many boilers have had multicyclones installed in past years in attempts to meet code requirements.

It is suggested that a particle size analysis be taken both ahead of and behind any existing cyclone collector before design selection of a wet scrubber. If the particle size distribution shifts strongly (more than 25% increase in particles less than 5μ) when a multicyclone is used, it is suggested that the cyclone be removed in favor of a larger device. The scrubber pressure drop requirements can be significantly less in many instances, prompting a more rapid payback of the dry collector cost and reducing abrasive wear in this device.

On new installations, the designer should consider the cyclone collector merely as a particle separator intended for > 50-μ species. It should not be relied on, in itself, to meet particulate codes. It reduces the scrubber inlet grain loading on species which could cause abrasive wear in the scrubber, while minimizing energy-consuming size reduction.

An accurate inlet grain loading to the scrubber should be determined. On existing systems, the flue ash handling system should be studied for leakage and other detriments to operation. It must function well if it is to be compatible with the scrubber/absorber system.

Wet scrubbers are required to reduce the boiler dust loading to code levels, precondition the gas stream for subsequent SO_2 removal, permit draft adjustments on the boiler, and to perform these tasks in as simple a manner as possible. Designers typically select venturi scrubbers for this application; however, a number of tray and cyclonic units have been used (the former on oil tanker boiler inert gas systems, and the latter on waste fuel boilers). The venturi scrubbers are usually adjustable-throat designs of from 6 to up to 20 in. w.c. depending on the application. As in other applications, experience and data (particulate size primarily) are used to determine the pressure drop. The designer should acquire data relative to the following items:

(1) Inlet particulate size analysis
(2) Ultimate analysis of the fuel
(3) Analysis of the makeup water
(4) Boiler design CO_2 outlet conditions
(5) Boiler turndown requirement
(6) Type of fuel combustor
(7) Layout of intended location
(8) Site plan of plant and surrounding area

Item (1) is used to determine a first estimate of the pressure drop as demonstrated in Section 1.3.

Item (2) is needed to project the possibility of submicron species. The ultimate analysis is studied for metallic species which could create submicron particulate upon condensation. These items are vanadium, silicon, sodium, magnesium and other metals which might exist in the fuel as organometallic compounds. One does not believe the particle size analysis until a thorough investigation of the fuel is carried out. Burning a high vanadium content No. 6 oil can add 3–6 in. w.c. to a venturi pressure drop. A projection should be made of alternative fuel sources which may, in the future, be used so that the scrubber and fan may be adequately sized.

Item (3) is important because high chloride or dissolved salt quantities in makeup water can destroy even stainless steel scrubbers. In boiler scrubbing systems, the chemical mix is an unsavory one. Acidic sulfur and chloride species are typically mixed with sodium or calcium alkalis at elevated temperatures — a perfect environment for stress corrosion cracking and pitting. Also, the introduction of sodium-containing makeup water at the inlet (high-temperature) zone of the scrubber may promote spray drying of these compounds, creating a submicron fume. In many instances it is better to mix decanted water from the flue gas desulfurization (FGD) system than to use fresh water.

Item (4) is required to determine the corrected outlet loading when the flue gases are adjusted for CO_2 content. Most codes stipulate 12 % CO_2 correction. Therefore, systems whose average flue gas CO_2 is lower will require lower outlet loadings. Thus it is imperative to have good burner control for maximization of combustion performance and CO_2 loadings.

Item (5), boiler turndown requirement, sets the range of operation of the scrubber. One must know if the flue CO_2 loading will also change on turndown so that pressure drops can be calculated for lowest and highest combustion rate. Also, the intended duty (base loaded, peaking, swing) of the boiler must be known. Automatic draft controls derive their engineering inputs from the intended duty. Drain back provisions on peaking or swing duty boilers are important, determining freeboard requirements of recirculation tanks and holdup facilities, as well as liquid seals.

Item (6), the type of fuel combustor, is obviously important. High-turbulence pulverized coal burners create excellent heat release and very fine particulate. Stoker-fired boilers tend to run at higher excess air levels, yet have somewhat reduced particle loadings. These data help put the given particle size analysis in perspective. Unduly high amounts of submicron particulate from stoker-fired boilers would draw attention to the particular fuel used during the test. Sometimes adjustments can more readily be made in the fuel source than in the cleanup equipment.

Items (7) and *(8)* help the designer select the proper scrubber to suit the plant's physical constraints. He may wish to scrub particulate in one vessel and use an entirely different vessel for SO_2 control. He may be forced to combine them into one unit. He may be able to place his waste handling system directly under the scrubber, or be forced to pump it to another facility. The breeching may be simply connected to the scrubber or require extensive modifications. The scrubber stack may leave the building in the lee of a taller stack—a poor arrangement for discharge of saturated gases. Or, it may be able to enter the existing stack, saving complexity and money. The site places important constraints on all systems designs, even where a new facility is being contemplated starting from the ground up. The surrounding area is important in selecting the required stack height for proper dispersion.

3.8.2 Gas Absorption

The absorption of SO_2 is the predominant function of absorbers applied to industrial boilers. There exist a variety of gas absorbers available to the applications engineer, deriving their use from the type of sludge disposal system to be used. These sludge systems are separated into two categories:

(1) *Recoverable or regenerable:* the absorbed SO_2 is chemically treated so that a reusable, if not resalable, product is created. Some products are gypsum for use in wallboard, elemental sulfur, sulfuric acid and sulfur dioxide.

(2) *Throwaway:* these would be better described as "storage" systems since the by-products must be further handled, perhaps once when produced (in the case of a settling pond) and years later when the storage area must be enlarged or excavated. These waste products have little or no market value. The most prevalent throwaway material is calcium sulfate.

These systems may be cyclic in the absorption circuit yet be batch in the sludge handling circuit. The gas absorber becomes a minor part of the entire system, and the wet chemistry becomes most important.

Several commercially available regenerative FGD systems are listed in Table 3.10. Generally speaking, regenerative systems require thorough particulate separation from the gas stream. These systems utilize closed-loop

TABLE 3.10. Regenerative FGD Systems.

Process	Reagent, Primary	Product
Double alkali (dual alkali)	Sodium hydroxide	Calcium sulfate (caustic returned)
Dilute sulfuric acid	Sulfuric acid	Calcium sulfate (acid returned)
Wellman-Lord	Sodium sulfite	SO_2
Magnesium oxide	Magnesium oxide	SO_2
Citrate	Sodium citrate	Sulfur

venturi scrubbers as particulate collectors or precipitators or baghouses. Required efficiencies are over 98% in many cases. Even small amounts of abrasive particulate can cause operational problems in the regeneration system. Venturi scrubbers must be 6–15 in. w.c. on coal-fired boilers for proper removal efficiency.

Retained fly ash will build up in regenerative flue gas desulfurization systems, reducing the operating life of the system itself and detracting from the product quality. Many pilot FGD systems of a regenerative nature perform well, only to have full-scale systems become operational nightmares. Large pilot facilities (10% or more of expected flue gas volume) are suggested.

Absorbers for regenerative systems can have smaller openings than their counterparts experiencing a larger mass flow of particulate. These gas absorbers may be tray towers, spray towers, bubble cap tray towers, packed towers or other devices described earlier. These absorbers, in general, operate with dilute scrubbing liquids since the end products desired have, in many instances, their own vapor pressure effects. Concentrated solutions may evolve other contaminants through a stripping action. Many systems start with sodium carbonate, sodium hydroxide or other soluble alkali which acts as a carrier of the sulfate or sulfite radical for further liquid-phase chemical operations. Thus, the absorber circuit may be cyclical, containing its own recirculation and makeup alkali system, while the regeneration stage may be batch. Extensive texts are available on the chemistry of these various systems for reader reference.

The absorber of preference is a spray tower or similar device wherein the vessel exposes the greatest surface area per unit volume to the gas stream. Devices with weirs, caps, baffles, etc., are, though effective, not as desirable given their higher pressure drops and internal complexity. Liquid-to-gas ratios (L/G) are high (e.g., 25–100 gal/1000 acfm) given the low concentrations of reagents. Many reactions are slow in execution and therefore require retention tanks. When this is the case, the absorber should discharge directly into the retention tank, or by the shortest path. Horizontal runs are avoided, even when soluble alkali is used, so as to reduce scale deposition or settling of residual fly ash. Pumps should be avoided between the absorber and the retention tank because they can present operational problems in both devices.

Retention tanks should accommodate the possibility of settling of solids, should provide for agitation and blow-back. Tanks should be fully drainable, if possible, and not designed only for pump-out. Instrumentation, such as pH probes, should be protected by stilling wells that are exposed to constant agitation. These probes must be fully accessible and, obviously, spares should be stocked. Tanks mounted indoors will evaporate contaminant and reagent along with steam; therefore covered, vented tanks are suggested. These covers should be fully removable, if possible, for worker access. Rubber-lined tanks should include shop-installed welding clips, reducing the chance of accidental welding onto the surface of previously lined vessels. Linings should be checked for thickness upon installation at a spot which is accessible for future checks to determine lining wear or degradation. Heat tracing, if used, should be externally available to the operating engineer.

Since the absorbing liquid typically returns from processes wherein heat is lost, the absorber typically condenses water vapor (especially if installed after a wet scrubber for particulate removal), causing a net increase in water vapor to the absorber circuit. This dilutes the reagent, and must be compensated for in the circuit water balance. Many times this forces a bleed of some valuable alkali from the absorber so as to physically maintain water levels. When this bleed cannot be utilized as shower water, reagent dilution or mix water, design scopes must include some means of water treatment for this flow. One method involves sending excess water to the particulate removal circuit as makeup. If this flow contains soluble alkali, one must anticipate the effects on the presence of dissolved salts in the fly ash disposal system. This water buildup can be equally important if a dry primary collector is used, especially where counterflow absorbers are used (the coolest scrubbing liquid sees gases last). The evaporation rate is reduced from the theoretical amount unless the balance includes the condensation of water vapor in the absorber.

Throwaway systems are easier to design if one neglects the ultimate disposal of the waste liquor or sludge. Many times, the sludge may be disposed of in landfills or ponds, prompting an engineering decision relating more to materials handling and less to chemistry.

The ultimate disposal of the sludge determines the design of the scrubbing system. Soluble alkali systems which absorb SO_2 using sodium hydroxide or sodium carbonate are well suited to systems where the resulting sodium sulfite and sulfate can be reused, such as in the kraft pulping industry. These scrubbers are simple in design, using a low- (6–12 in.) energy venturi scrubber with adjustable throat followed by a tray absorber or packed tower. Some units do not even use the absorber for 80% SO_2 removal, given the high activity of caustic on SO_2. These units operate on combined-loop recirculation systems with makeup to the venturi sump for caustic and fresh water makeup to the venturi headers to control evaporative losses. Figure 3.18, from Dow Chemical [3], shows the effects of recycle pH on SO_2 absorption.

Fly ash is settled out and the supernatant liquor returned to process. This is also the first stage of the dual alkali system. In the dual alkali system, this liquid flow is then slaked with lime to create calcium sulfate solids and a caustic supernatant which is returned to the first stage. Thus it is a throw-

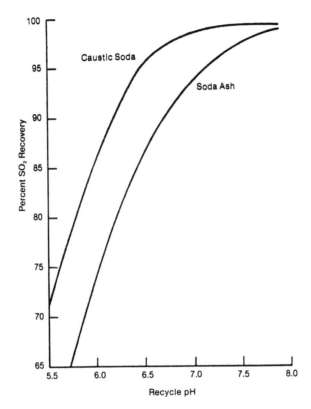

FIGURE 3.18. Sulfur dioxide recovery vs. recycle pH (source: Dow Chemical USA [3]).

away system (calcium sulfate to landfill) and a regenerative system (caustic returned to the scrubber) in one. Unfortunately, cross migration of insoluble compounds can occur, yielding operational problems in the scrubber (scale, erosion, etc.). Once again, condensation in the absorber can occur, requiring a liquid balance for each particular device. To avoid spray drying of caustic at the inlet of the venturi, makeup water alone should be introduced into the hot sections of the venturi with returned caustic entering once the gas temperature is below 400°F. Level controls in the absorber prompt increases in liquor additions to the venturi circuit. This feedback control is a common remedy for liquid level "hunting" in dual stage scrubbers. One uses the sensor in the least fluctuating device (usually the absorber since it is downstream from areas of evaporation and high carryover) feeding back to control valves on the most sensitive device (usually the scrubber). This applies to level control, pH and conductivity measurements.

3.8.3 Scrubbing System Components

Throwaway lime or limestone systems have five major areas, each requiring a mass balance:

(1) Slurry preparation
(2) Absorber
(3) Recycle/retention tank
(4) Wash tank
(5) Waste disposal

Lime or limestone is thoroughly mixed with water and recycled supernatant from the system's waste disposal process. Agitators blend this slurry to a weight solids content of 15%. Rubber-line centrifugal pumps distribute this slurry on a continuous flow loop to the absorber, a valve actuated by pH instrumentation controlling slurry addition, and a density system controlling dilution.

The absorber may be a tray tower, but the most prevalent design is an open spray tower. These devices rely on a spray of hydraulically atomized scrubbing liquid to provide the surface area necessary for gas-to-liquid exchange. Prehumidification sprays are sometimes used, incorporating water, to create sulfurous acid when then reacts quickly with the alkali spray. If insufficient alkali is present, the reaction can produce hard and soft scale deposits. Therefore, the scrubbing liquid balance must avoid saturating the liquid in any zone of the absorber. This is accomplished by returning supernatant from the clarification step of the waste disposal system and adding makeup water to the chevron droplet eliminator or wash tank. There is always alkali available for reaction with SO_2.

The absorber is typically a vertical or horizontal chamber with spray headers. Sometimes target devices are used to increase turbulence in the absorber. Some typical design details for various components of an absorber are listed in Table 3.11.

The recycle system is one of the most important parts of the lime/limestone absorption system. It is estimated that chemical reactions to sulfate may take up to 10 min to go to completion. This requires adequate holdup to avoid sending partially reacted scrubbing liquid back to the absorber where it may scale. Thus, the agitated recycle tanks can contain up to 10 min retention time, sometimes many thousands of gallons. This is prompted by the high L/G, which in turn is prompted by the need to create absorption surface area via hydraulic forces. It is known that the greatest surface area per unit volume exists a short distance from a spray nozzle, decreasing from there out from the nozzle. New devices, such as the Catenary Grid Scrubber® (patented), Figure 3.19, use grids of widely space wire to act as

TABLE 3.11. SO₂ Absorber Application Guide.

Specify: Spray tower vertical velocity, 8–10 ft/sec
Header velocities, 6–8 ft/sec
Chevron mist eliminators face velocity, 550 ft/min based on open area
(800 ft/min max.)
Pressure drop, 0.5–1.5 in. w.c.
Chemical and Liquid Requirements (for limestone scrubber):

Emission Rate	Mol Limestone/Mol SO₂	Required L/G, gal/1000 acf Saturated Gal
1.5 lb SO₂/10⁶ Btu	0.65–0.75	50
1.0 lb SO₂/10⁶ Btu	0.75–0.8	80
0.5 lb SO₂/10⁶ Btu	0.9–0.95	90

targets for rebreaking of the scrubbing liquid. This increases the aggregate surface area of the scrubbing liquid, and increases absorption per unit volume without using higher header pressures or other less efficient atomization techniques.

Waste handling systems include dewatering systems (vacuum filters, clarifiers or combined device systems), holdup tanks, and decant water return. This decanted or filtered water, commonly called supernatant, is returned for reuse as shower water or makeup water. A holdup tank is required since its flow can be discontinuous. It also assists in starting up the system, permitting the waste disposal system to operate without the absorber circuitry in operation.

Sometimes the waste is sent to settling ponds. This mixture of dissolved solids (low quantities of chlorinated metallics), suspended solids (calcium sulfate and sulfite), and water is sometimes blended with the fly ash recovered from the particulate removal device. Adequate facilities must be made for this discharge. Therein lies the argument against throwaway systems. The expense and complexity of the solids disposal system must be investigated as a significant portion of the investment.

Soluble alkali systems, those using sodium hydroxide or sodium carbonate or other similar alkalis, are throwaway systems of "first the good news, then the bad news" type. The good news is that the use of a soluble alkali reduces the effects of scaling on the scrubber absorption surfaces. This improves the on-stream reliability of the system and reduces off-line maintenance expenses. The bad news is that the waste liquor contains a mixture of sodium bisulfite and sulfate, along with (especially in coal burning applications where the fuel contains chlorides) residual halogens. Typical chemical composition of such a system blowdown as a function of pH is given in

Figure 3.20. These items are leachable into groundwater, requiring further treatment prior to disposal. Hence, we have the dual alkali systems wherein a soluble alkali absorber is mated with a water treatment system that typically slakes the soluble waste stream with lime or limestone, creating an insoluble waste product. This operation requires exacting attention to the wash circuitry in the waste treatment area. Obviously, excessive water consumption in an attempt to thoroughly reduce the leachable salts in the waste slurry is undesirable. In general, a double alkali system requires greater

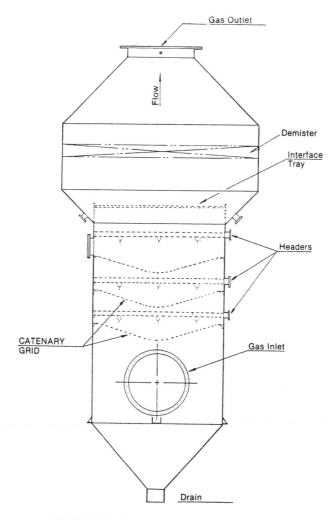

FIGURE 3.19. Catenary Grid Absorber® (patented).

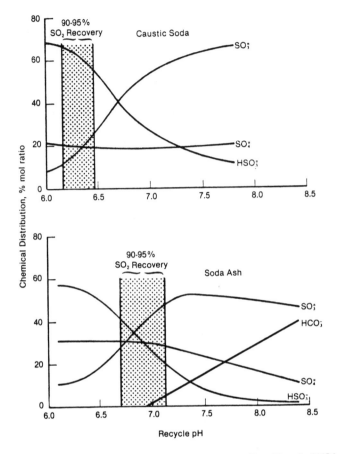

FIGURE 3.20. Blowdown composition vs. pH (source: Dow Chemical USA [3]).

wash flow rates to achieve the same waste sludge conditions as a direct lime/limestone system. With the scaling problems overcome in lime/limestone systems, given better internal loop chemistry, these systems are generally preferred over soluble alkali systems unless the waste liquor can be used in the process directly without further treatment.

Given the higher reactivity of the soluble alkali systems, the absorbers are simpler. One finds tray towers, spray towers, bubblecap trays, wetted chevrons, packed towers, air washers and low-energy venturi scrubbers all performing well on this application, as shown in Table 3.12. Note the much lower L/G requirement than the limestone values in Table 3.11.

Caustic is premixed with water and then fed into the absorber sump. The caustic should not be fed to the inlet of the scrubber unless one wishes to spray dry caustic. It should be permitted to mix and react with the absorbed SO_2 in the sump, and be further mixed by the recirculation pump before being sent to the absorber media. Makeup water should be directed at the bottom of the mist eliminator device or at the entry to the scrubber, serving to cool and saturate the gases. The scrubbing liquor should be controlled by a pH controller mounted in the pump discharge bypass or in a stilling well in the scrubber sump so that it measures the internal loop (not the bleed) pH. A worthwhile addition is a sample tap in the recycle line to take samples for sulfite/bisulfite analysis. Tables 3.13 and 3.14 show the basic chemistry of the absorption of sulfur dioxide and its conversion to salts when using lime, limestone, sodium hydroxide, and sodium carbonate. These reactions are based upon a stoichiometry of 1.0; therefore, in actual operating systems chemical reagents in excess of these figures may be needed. As mentioned previously, designers balance the removal efficiency requirements versus operating constraints (such as the propensity of the scrubbing liquid to produce scale that may plug the absorption device) in selecting the proper operating point. These tables will provide the reader with guidelines, however, as to the approximate chemical consumption for these reagents.

3.9 ODOR CONTROL SCRUBBERS

3.9.1 Introduction

Odors consist of particles that are odorous or contain adsorbed odors and/or odorous gases. Scrubbers are capable of removing both gases and particles, and can be useful for odor control. Table 3.15 shows a crude listing of various dry odor control techniques and the approximate times required for 90% reduction of the odor. This is expressed as odor units (o.u.). Time for wet scrubbing odor control includes time for diffusion and absorption plus chemical reaction.

TABLE 3.12. Sodium-Based Absorption Application Guide.

Parameter	Value		
SO_2 removal, % efficiency	90	94	96
Recycle:			
lb NaOH/lb SO_2	0.85–1.0	0.9–1.0	1.0–1.1
lb Na_2Co_3/lb SO_2	1.85–2.2	2.1–2.4	2.2–2.6
pH (internal loop)	6.5–6.8	6.8–7.0	7.0–7.2
Mol sulfite/bisulfite	0.5	0.6–0.7	1.0
Absorber, number of theoretical trays	1.5	2.0	2.5
Tray tower (L/G), gal/1000 acf			
saturated gas	10	10	10

TABLE 3.13. Sulfur Dioxide Absorption Chemistry.

Using sodium hydroxide (NaOH):

$$2NaOH \;+\; SO_2 \;\rightarrow\; Na_2SO_3 \;+\; H_2O$$

| 80 | + | 64 | → | 126 | + | 18 |
| m.w. | | m.w. | | m.w. | | m.w. |

$$80/64 \;=\; 1.25 \text{ lb NaOH per lb } SO_2$$

and

$$NaOH \;+\; So_2 \;\rightarrow\; NaHSO_3$$

| 40 | + | 64 | → | 104 |
| m.w. | | m.w. | | m.w. |

Using sodium carbonate (Na₂CO₃):

$$Na_2CO_3 \;+\; SO_2 \;\rightarrow\; Na_2SO_3 \;+\; CO_2$$

| 106 | + | 64 | → | 126 | + | 44 |
| m.w. | | m.w. | | m.w. | | m.w. |

$$106/64 \;=\; 1.66 \text{ lb } Na_2CO_3 \text{ per lb } SO_2$$

and

$$Na_2CO_3 \;+\; H_2O \;+\; 2SO_2 \;\rightarrow\; 2NaHSO_3 \;+\; CO_2$$

| 106 | + | 18 | + | 128 | → | 208 | + | 44 |
| m.w. | | m.w. | | m.w. | | m.w. | | m.w. |

$$106/28 \;=\; 0.83 \text{ lb } Na_2CO_3 \text{ per lb } SO_2$$

TABLE 3.14. *Sulfur Dioxide Absorption Chemistry Using Lime or Limestone.*

Using lime ($Ca(OH)_2$)

absorption

SO_2	+	H_2	$\rightarrow$	H_2SO_3		
64	+	18	$\rightarrow$	82		
m.w.		m.w.		m.w.		

H_2SO_3	+	$Ca(OH)_2$	$\rightarrow$	$CaSO_3$	+	$2H_2O$
82	+	74	$\rightarrow$	120	+	36
m.w.		m.w.		m.w.		m.w.

also

H_2SO_3 + $CaSO_3$ $\rightarrow$ $Ca(HSO_3)_2$

$Ca(HSO_3)_2$ + $Ca(OH)_2$ $\rightarrow$ $2CaSO_3$ + $2H_2O$

oxidation

$2CaSO_3$ + O_2 $\rightarrow$ $2CaSO_4$

pH 5.5 $\rightarrow$ 6.8

overall

SO_2	+	$Ca(OH)_2$	$\rightarrow$	$CaSO_3$	+	H_2O
64	+	74	$\rightarrow$	120	+	18
m.w.		m.w.		m.w.		m.w.

$74/64 = 1.15$ lb lime per lb SO_2

Using limestone ($CaCO_3$):

$CaCO_3$	+	SO_2	$\rightarrow$	$CaSO_3$	+	CO_2
100	+	64	$\rightarrow$	120	+	44
m.w.		m.w.		m.w.		m.w.

$100/64 = 1.49$ lb $CaCO_3$ per lb SO_2

pH = 5.8–7.0

oxidation same as lime

TABLE 3.15. Comparison of Odor Control Techniques by Dry Processes.

Process	Time Required for Odor Control, >90%	Dependence on Chemical Nature of Odorants
Adsorption	~0.04 sec	Not dependent
Flame oxidation	~0.4 sec	Moderately dependent
Absorption	Varies, e.g., typical residence times = ~1 sec in tower = ~0.5 sec in venturi	Dependent
Dry gas-gas or gas-liquid phase oxidation	Slow = several sec	Strongly dependent, but oxidation rate can be greatly speeded up by catalysts

3.9.2 Chemical Oxidants

Chemical oxidation of gas streams to destroy odor is a preferred technique. The type of contact can be any combination of one or more of gas-gas, gas-liquid or gas-solid. Oxidizing gases used in gas-gas contacting systems include chlorine and ozone. Both of these are toxic gases (and as such are pollutants), so use of these gases must be confined and controlled to prevent discharge of any unreacted gas. Chlorine is usually used in conjunction with a wet scrubbing system because of its toxicity, residual odor and corrosiveness, although other substances are also used. Ozonated scrubbing liquid is a recent variation and results in higher ozone utilization than in gas ozonation and a reduction in oxidation time (e.g., from 30 sec in a dry system to 3 sec for a wet system for one specific odor).

Other oxidants can be used in wet scrubbers. This includes chemicals such as potassium permanganate, chlorine, sodium hypochlorite, chlorine dioxide, hydrogen peroxide, sulfuric acid or caustic. Chlorine dioxide should not be used in the presence of ammonia, methane, hydrogen sulfide or hydrogen phosphide, as it could result in explosive mixtures. Chlorine dioxide works well in a solution of any pH.

Chlorine requires an acid solution to be an effective oxidizer. Sodium hypochlorite produces an active chloride compound, hypochlorous acid (HOCl), which acts similarly to chlorine in an acid solution. Chlorine is not effective against methyl or ethyl amines, mercaptans or sulfides.

Potassium permanganate acts as an oxidizing agent to odorous air streams and works best when the solution pH is near 8–9.5. Buffers, such as sodium carbonate or bicarbonate or borax, are added to try to stabilize pH. Absorbed CO_2 and oxidation products would otherwise rapidly lower pH. Potassium permanganate is considered to be about the most versatile oxidant in that more odors can be treated by it, but it is not the oxidant for any particular class of odorants. Expended potassium permanganate is in the form of insoluble magnesium dioxide. This provides a gas-solid surface area where adsorption can take place. Optimum potassium permanganate concentration for oxidation is 1–2%. Permanganate solutions can oxidize H_2S, organic sulfide and aldehydes, aliphatic amines, mercaptans, phenols and heterocyclic compounds. Potassium permanganate must be dissolved in warm water, which could be troublesome. The MnO_2 formed by the oxidation reaction can plug scrubbers, and periodic cleaning is thus required (e.g., weekly) to restore original system ΔP. Cleaning is accomplished with a sodium bisulfite solution. Cleaning times for this can vary from 1 hr on up, depending on the amount of deposit. Metal corrosion rates can be 0.00002 in./hr or more during cleaning even when corrosion inhibitor has been used.

Hydrogen peroxide is a long known oxidant which leaves no contaminant after decomposing. In alkaline solutions it will oxidize sulfides, thiosulfates and polythionates. Sulfites are oxidized in acid solutions.

Oxidizing acids such as sulfuric acid are used alone or with solid oxidants to control odors. Usually they are followed by a base to neutralize mists entrained in the vapors. Although the base calcium hydroxide is not an oxidant, it appears to be an effective reagent for controlling mercaptan odors.

3.9.3 Scrubbing Systems

Minimizing oxidant chemical addition is important to minimizing operating costs. Therefore, it is desirable to precondition the odorous gas stream by first removing the particulates, as is noted in many of the ozone treatment examples. This can be done in a mechanical collector or condenser, but many operations prefer to use a low-efficiency wet scrubber as the first removal stage. Often some chemicals are added in this stage. A second-stage scrubber nearly always uses chemicals in the scrubbing slurry. Effective mist eliminators must follow each scrubber to remove entrained liquid. Schematics of such systems are shown in Figures 3.21 and 3.22. Figure 3.23 shows a baghouse-wet scrubber combination.

In most scrubbing systems reagent concentration is not critical to the effectiveness of odor removal. It is of course necesary to have adequate reagent available to react with the absorbed odor vapors. Permanganate and hypochlorite seem to be the most concentration-dependent and show a slight positive effect for reagent concentration on odor adsorption rate (i.e., higher concentrations are slightly more effective). Hydrogen peroxide is also an effective odor control scrubbant. Such a system is shown in Figure 3.24.

Control of the chemicals is required in these systems and a procedure for

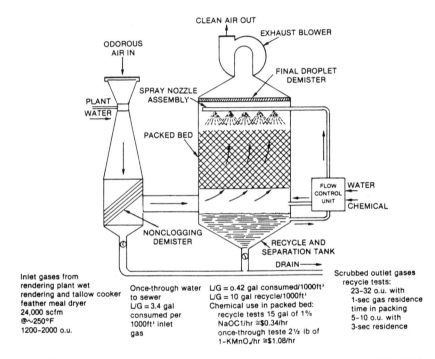

CLEAN AIR OUT

EXHAUST BLOWER

ODOROUS
AIR IN

FINAL DROPLET
DEMISTER

SPRAY NOZZLE
ASSEMBLY

PLANT
WATER

PACKED BED

FLOW
CONTROL
UNIT

WATER

CHEMICAL

NONCLOGGING
DEMISTER

RECYCLE AND
SEPARATION TANK

DRAIN →

Inlet gases from
rendering plant wet
rendering and tallow cooker
feather meal dryer
24,000 scfm
@~250°F
1200-2000 o.u.

Once-through water
to sewer
L/G = 3.4 gal
consumed per
1000ft³ inlet
gas

L/G = o.42 gal consumed/1000ft³
L/G = 10 gal recycle/1000ft³
Chemical use in packed bed:
  recycle tests 15 gal of 1%
  NaOCl/hr ≅$0.34/hr
  once-through teste 2½ lb of
  1-KMnO₄/hr ≅$1.08/hr

Scrubbed outlet gases
recycle tests:
  23-32 o.u. with
  1-sec gas residence
  time in packing
  5-10 o.u. with
  3-sec residence

FIGURE 3.21. Wet scrubbing system using venturi first stage and countercurrent packed tower for odor control.

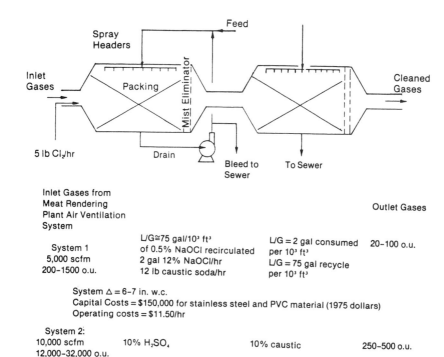

FIGURE 3.22. Wet scrubbing odor control system using horizontal cross-flow scrubbers in series.

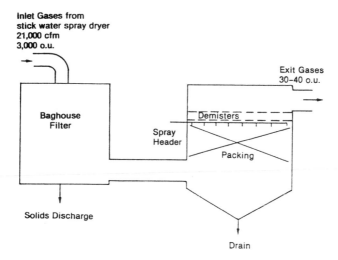

FIGURE 3.23. Baghouse filter plus packed tower scrubber for odor control.

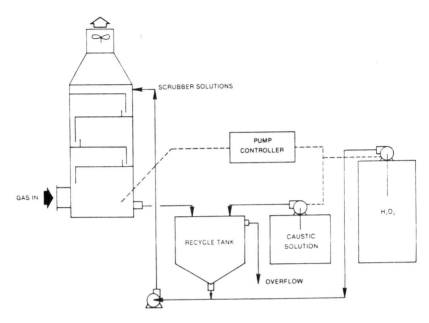

FIGURE 3.24. Hydrogen peroxide scrubber for odor control (courtesy Peabody Engineering Corporation).

waste neutralization may be necessary. Figure 3.25 shows a pH analyzer and probe that could be part of a chemical control system. Figure 3.26 is an example pH neutralization system. An application guide for odor control systems is presented in Table 3.16.

3.10 HYBRID SCRUBBERS

When more than one separation technology is required to remove pollutants from a gas stream, designers often have to combine different wet scrubber designs into one scrubber housing or into connected units. For the purposes of this chapter, we will call these designs "hybrid" scrubbers.

An example of this type scrubber would be the problem of absorbing acid gases from a gas stream containing acid aerosols. As discussed in previous chapters, the gas absorption problem could be addressed using a packed tower scrubber, tray tower scrubber, spray scrubber, or fluidized bed scrubber. The aerosols, however, are not easily collected as a gas. Since an aerosol is a liquid *particle,* they are collected by inertial, electrostatic, or Brownian diffusion techniques.

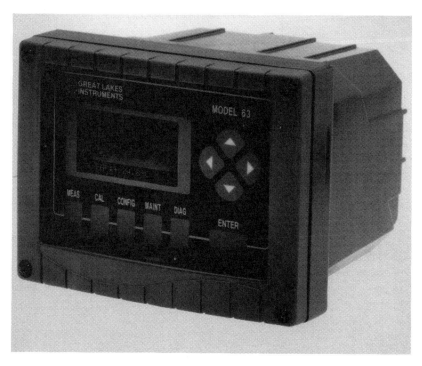

FIGURE 3.25a. Chemical control using pH analyzers (courtesy Great Lakes Instruments).

FIGURE 3.25b. Chemical control using pH analyzers. pH probe assembly (courtesy Great Lakes Instruments).

150

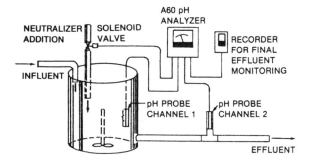

FIGURE 3.26. Dual-channel pH neutralization system (courtesy Great Lakes Instruments).

The acid gas/aerosol mixture problem is common in the semiconductor manufacturing industry where strong, sometimes heated, gases are used. Heated nitric acid and aqua regia tanks, for example, are used in the process. Mysterious "plumes" are visible from the system vents that are not equipped with the proper control equipment. The vents from these tanks contain gaseous acids. When cooled or when water vapor is added, a por-

TABLE 3.16. Odor Control Systems Application Guide.

Typical System: Dual- or triple-stage gas absorbers.

Type of Absorber: Packed-tower of tray tower, grid scrubber, cross-flow washers, wetted web filters.

Technique: Extend liquid surface using media or sprays or both. Utilize neutralization followed by oxidation wet chemical techniques. Afterscrubber sometimes used.

Control: pH control on neutralization stage and ORP or specific-ion electrode ion oxidation stage.

Liquid Circuit: Neutralization chemicals are prepared on a batch basis and bled on demand into the neutralization stage. Oxidant is fed from the reagent tank or from batch preparation tank to the oxidant scrubber. A sump typically provides added residence time for oxidant reaction to go to completion. Since an excess of oxidant is used to increase the driving force in the reaction, a residual oxidant odor can result. A water afterscrubber sometimes is required.

Particulate: Particulate must be removed before odor control.

Materials: To suit reagents. Typically fiberglass construction with PVC or polypropylene internals. Teflon® spray nozzles frequently are used.

Mist Elimination: Mesh-type high-efficiency droplet eliminators are required.

Test Method: For specific contaminant or via odor panel.

Special Considerations: Many odor problems are under the jurisdiction of the local department of health, which has specific testing methods. Local regulations always should be consulted.

tion of these acids form acid aerosols. Figure 3.27 shows the typical pathway for this aerosol formation. The introduction of water vapor along with cooling through contact with a colder air stream or other lower temperature stream can cause the aerosol to form.

A solution to this problem is the use of "hybrid" scrubber incorporating a counterflow, pH adjusted packed tower for gaseous acid absorption followed by a fiberbed filter for Brownian motion aerosol capture. In Figure 3.28, the system components are described. The packed tower is designed to operate at a pH of 9–10 using caustic on automatic pH adjustment using KOH (potassium hydroxide), and the fiberbed is configured to remove the acid aerosol. The drain water from the fiberbed usually has a pH of less than 2, indicating that the aerosol has been captured on the fiberbed's microfibers, has accumulated sufficiently to drain, and has been converted to a liquid.

Another example occurs in the manufacture of printed circuit boards. The boards are dipped in hot (450°F) oil to level the solder placed on the boards in a previous manufacturing step. When the boards are removed from the oil, oil smoke and odor is evolved. Electrostatic precipitators had been used that required constant maintenance to remove accumulated oil residue. Figure 3.29 shows the basic components of the "hybrid" scrubber that solved the emission problem. It consists of a counterflow Catenary Grid Scrubber™ using pH adjusted scrubbing solution plus hydrogen peroxide for odor reduction, followed by a fiberbed filter. The scrubber acts as an absorber *and* condenser. It absorbs the water soluble odors and helps to condense the oil vapors so that they form collectable aerosols. The fiberbed filter collects the oil aerosols. Figure 3.30 is a photograph of the installation. The unit on the right is the wet scrubber, and the unit on the left is the fiberbed.

Another "hybrid" wet scrubber design would be the use of a scrubber after a "dry" scrubber, such as is used for SO_2 removal on power plants. In these systems, a spray dryer is usually followed by a baghouse (filter). An alkaline solution or slurry is introduced into the spray dryer and the acid gases are absorbed into the solution. The dried slurry, along with pollutant fly ash, is captured in the baghouse. Typical acid gas capture efficiency is 80–90% without a supplemental scrubber. What if 99% capture is required? A "hybrid" scrubber could be designed using a fluidized slurry or spray tower contactor located after the baghouse. If the scrubber is designed to recycle slurries of sufficiently high solids content (over 10% by weight) the slurry could be returned to the spray dryer for drying. Figure 3.31 shows the general layout of this type system. The key factor here is the ability of the wet scrubber to operate at high solids loadings. Packed towers and certain spray towers using nozzles or packing that could plug would not be useful for this application.

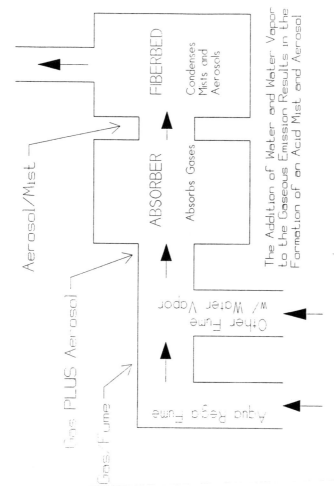

FIGURE 3.27. The addition of water and water vapor to the gaseous emission results in the formation of an acid mist and aerosol.

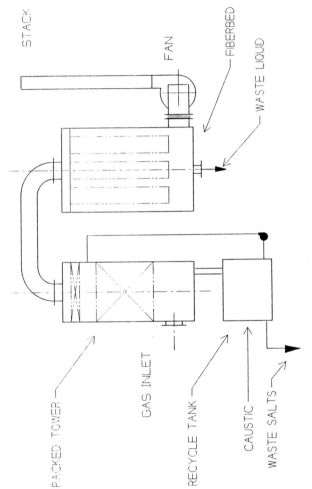

FIGURE 3.28. Two stage acid gas scrubber.

154

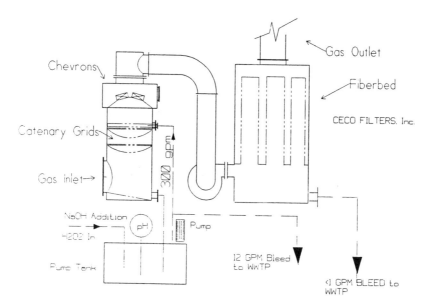

FIGURE 3.29. Multistage gas cleaning system—20,000 acfm design.

When the particle size distribution of a pollutant gas stream contains over 50% less than 1 μ, the energy consumption of venturi type scrubbers can, at first glance, appear prohibitive. As mentioned in previous chapters, a venturi scrubber's pressure drop requirement is directly related to the size particle one wishes to capture. As the particle becomes smaller, the venturi scrubber energy requirement increases.

Designers, however, have successfully used a technique called flux force condensation (FFC) scrubbing to "grow" these particles to a more manageable size. It is used on incinerators and other sources that generate submicron fumes and have an adiabatic saturation temperature of above about 140–145°F. FFC scrubbing basically duplicates in this hybrid scrubber design what Mother Nature does in our atmosphere. When water vapor in clouds condense, researchers have shown that the water vapor tends to condense on particles. When clouds are "seeded" to help make rain, particles are injected into the cloud to provide these sites around which the water vapor would like to condense.

Researchers such as Dr. Seymour Calvert and Dr. Shui Chow-Yung of Air Pollution Technology (Riverside, CA) studied this technique in the early 1970s funded by government grants. The concept itself dates back to the 1800s and was used by Wilson to describe the path of cosmic rays through his famous "Cloud Chamber" experiment. These chambers show the aero-

FIGURE 3.30. Installation (CECO fiberbed filter on left).

FIGURE 3.30 (continued). Installation (Catenary Grid Scrubber®, first stage).

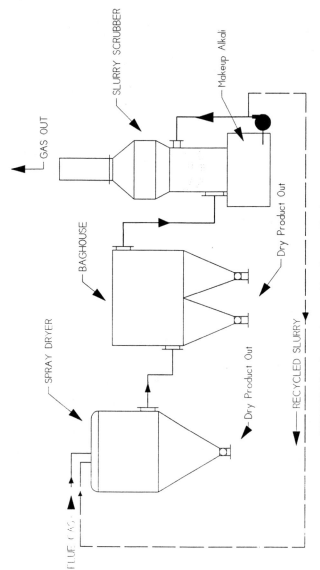

FIGURE 3.31. Spray dryer and baghouse with slurry scrubber.

158

sol traces that are formed by cosmic ray particles as they pass through a gaseous medium. They were and still are very popular at science fairs and public science museums around the country.

Figure 3.32 shows the basic components of an FFC system. It consists of a quencher that adiabatically saturates the gas stream with water vapor; a condenser/absorber that recirculates "cool" water, thereby causing the water vapor to condense onto the particles; and a particulate collection device (usually a venturi scrubber). The submicron particulate becomes covered with water vapor and "grows" aerodynamically to about 0.8 μ diameter. This doesn't seem like much, but the increased size greatly decreases the energy required for capture.

The venturi scrubber is typically followed by a high efficiency droplet eliminator and an induced draft fan. The heat that is extracted from the system can be recovered or dumped to atmosphere using a cooling tower or heat exchanger.

Key factors regarding FFC scrubbing involve:

(1) One must typically condense over twenty times more water vapor by weight than the amount of particulate. Also, it works best if the particulate loading is below 0.1–0.2 grs/dscf at the condenser/absorber inlet.

(2) It also works best if the saturation temperature exceeds 140–145°F, thus allowing enough cooling potential to utilize a standard cooling tower for heat rejection.

(3) The selection of the heat rejection device usually dictates the lowest temperature to which the gases can be cooled and therefore controls the amount of condensate that can be harnessed for particulate encapsulation. If a cooling tower is used, the lowest liquid temperature that can be used in the condenser/absorber is approximately 85–90°F. If a tubular heat exchanger is used (i.e., no evaporative cooling is available), the condenser/absorber water can only be cooled to 95–100°F in most areas of the country since the hottest daily temperature can approach those numbers.

(4) The venturi scrubber will see both larger particles and a smaller gas flow. The venturi pressure drops are typically 35–45 in. w.c.; however, the volume is usually half that of a conventionally applied venturi therefore the net energy input is about half.

(5) Cooling in the venturi is of little or no help. One must condense first, grow the particles, then remove them. The particulate collector must be efficient at removing 0.6–0.8 μ particles.

(6) Not enough water vapor? Saturation temperature too low? You can usually add steam ahead of the condenser/absorber to supplement the gas's water vapor content, then condense the mixture.

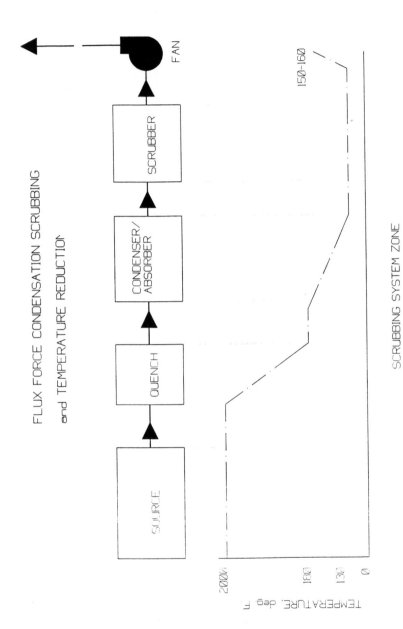

FIGURE 3.32. Flux force condensation scrubbing.

160

(7) The total energy consumption evaluation must include the costs of pumping the condenser absorber cooling water and the cooling fans (if any) on the heat exchanger. Typical net energy savings, however, are still over 25% and the system is usually smaller in size.

FFC scrubbers have been successfully used on hazardous waste incinerators, hospital waste incinerators, furnaces (smelting and remelting), reactors generating P_2O_5 and P_2S_5, and other fine particulate sources.

As a final example, we will discuss a multiple technology hybrid scrubber for remediating radioactive pollutants. This system controls the emission from a system that processes radioactive waste components. The block diagram in Figure 3.33 shows the major components. They include a prescrubber, a high-energy multiple throat venturi scrubber, an entrainment separator, a dehumidifier (condenser), a reheater, and finally an HEPA (high efficiency particulate air) filter. This system uses the separation tech-

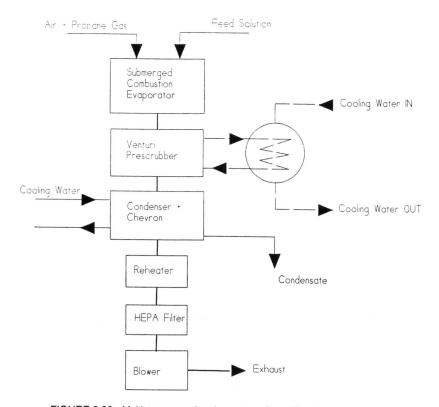

FIGURE 3.33. Multistage gas cleaning system for radioactive waste control.

nologies of absorption, condensation, impaction, diffusion, interception, and fiber filtration all in one system. Any one of these technologies alone would not solve the designer's problem. All of them had to be combined to provide process emissions low enough to meet stringent NRC (National Regulatory Commission) standards.

"Hybrid" wet scrubbers are a very important part of an air pollution control equipment designer's database. They are becoming increasingly important as air emissions regulations become tighter and more difficult to achieve using just one or two techniques.

3.11 WASTE MINIMIZATION USING WET SCRUBBING TECHNIQUES

One of the most interesting applications for wet scrubbers is their use to control incinerator emissions while producing a minimal amount of waste products. Sometimes we overlook the importance of minimizing the waste produced.

It is often intriguing to listen to the comments of young people when they learn of the discovery of a hazardous waste dump or of some company being cited for environmental offenses.

"Why did they do that?" they ask. "Didn't they know that what they were doing was wrong?"

As many environmental professionals can tell you, many of today's violators did nothing "wrong"; they just didn't do what was *right*. Codes and regulations in the past were less stringent than today. Illegal operations aside, many reputable firms operated in a manner that they felt best served the interests of their shareholders, their workers, and the community. Yes, they made compromises but they thought they were doing the *right* thing at the time. We have the luxury to make the judgement of what is "right" in today's context, assuming, perhaps as the violators did in their time that we truly know the *right* course of action when approaching environmental problems.

Our case in point is the creation of and composition of the by-products of modern air pollution control systems. Could we, in our current drive to solve air pollution problems, be merely creating a new set of environmental problems that may someday be looked upon with equal disdain? Wet scrubbers *can* be used to control pollution while producing a minimum amount of waste production. This section describes such a system.

3.11.1 The Ideal Gas Cleaning System

If we could build the "Ideal" gas cleaning system, it would:

(1) Remove contaminants from the gas stream at high efficiency
(2) Use minimal energy
(3) Be affordable to build and operate
(4) Be easy to maintain
(5) Be safe to operate
(6) Yield a waste product of minimum volume and minimum threat to the environment

Items (1)–(5) can be assumed to be the goals of many, if not all, air pollution control system designers. But item (6) may not be.

With landfill space becoming more scarce and more expensive, the need to produce a minimal volume of waste products should be self-evident. The impact of future cost escalations regarding waste disposal should make the volume of waste generated by a gas cleaning system a very important design parameter. It often isn't, however. Given the extensive handling (loading, trucking, unloading) of waste products from a gas cleaning system and potential worker exposure, you would think that the waste stability would be a very important factor, but it often isn't. If a landfill exists, and if someone will take it, some engineers feel the problem can be considered solved.

For example, in the treatment of large gas volumes from power boilers and certain incinerators, the use of spray dryers followed by a fabric filter collector has found widespread use. In these devices, the total particulate matter flow represents the minimum volume of waste product the gas cleaning system would produce. These "modern" systems then add to that volume absorbents such as lime [4], sodium bicarbonate, and other alkali to remove acidic components such as HCl and SO_2. These alkali materials are added at stoichiometric ratios of 1.1 to over 3.0, depending upon the removal efficiency required [5]. In some cases, bag precoat material is also added to improve dust cake removal and for filter bag protection.

To remove toxic organics, such as dioxins, even more material is added. Activated carbon, coke [6], and other adsorbents have been proposed or are in use for control of these toxic compounds. Since the mechanism of capture is *physical* adsorption, the captured compounds are, in most cases, not chemically altered or otherwise stabilized. Though they can be removed from the gas stream, the toxic organics can subsequently be stripped from the adsorbent.

For volatile heavy metals control, *even more additives* have been proposed. These involve the use of chemically impregnated carbon, various vapor injection systems, and other specific additives [7].

Table 3.17 describes a hypothetical spray dryer system from the point of view of the waste "generated" during the pollutant capture. Our description is on a mass basis, but a similar revealing comparison could be made on a

TABLE 3.17. Dry Scrubber with Additives, Waste Produced.

Contaminant	Emission Rate, kg/hr	Waste Ratio, kg/kg	Additives/Waste			Total Waste Produced
			$Ca(OH)_2/$ $CaCl_2$	Carbon	Impregnated A.C.	
Particulate	388	0	0	0	0	388
HCl	854	1.83	1563	0	0	1563
SO_2	2026	2.62	5308	0	0	5308
Hg	2.3	50	0	0	115	115
Dioxins	0.001	100	0	0.1000	0	0.100
Minimum waste = 3270 kg/hr						7374
Ratio of total waste produced to minimum waste produced =						2.25

Basis: 170,000 Nm^3/hr gas flow rate
2000 ppmv SO_2, SR = 1.4
1500 ppmv HCl, SR = 1.2
0.005 mg/Nm^3 Hg, 2% wt. carbon capacity at bleed through
40 ng/Nm^3 dioxins, 1% wt. carbon capacity at bleed through
Assume 100% capture of contaminants.

164

volume basis. It can be seen that the "dry" collection system produces far greater quantities of waste product than would have been produced if all of the pollutants had been removed without or at least with minimal additives. For our analysis, we used approximate inlet loadings for a municipal solid waste incinerator as reported in Reference [8]. However, the general trends are applicable to many similar combustion systems.

The predominant controlling parameter for the generation of the incinerator scrubber waste is not the control of particulate but the acid gas capture. Improving the efficiency of acid gas capture using techniques that produce less waste material per kilogram of contaminant can yield significant results in waste minimization. Properly applied wet scrubbing techniques can accomplish this goal.

How might item (6) in our list of parameters for the "ideal" system be met?

We would need to remove the pollutants, react them with a minimal quantity of reagents, and render them as stable as possible. Reagent selection and system design would be focused not just on pollutant removal but on final destruction and/or stabilization.

This would require a wholly different system. To reduce the quantity of chemicals used and to reduce the volume of waste products generated, we would need to break the gas cleaning problem down into specific components based upon the needed chemical reactions to meet the requirements of item (6). One such system, but in no way the only system, is shown as an example.

3.11.1.1 HYBRID SCRUBBING SYSTEM

One possible system to remove pollutants while producing a minimum waste volume would be a Hybrid Scrubbing System. In this system, different technologies are used to capture the pollutants and assist in their neutralization, stabilization and/or destruction.

Figure 3.34 shows this scrubbing system in its various components.

3.11.1.1.1 Particulate Control

The control of fine particulate using electrostatic, filtration, and wet scrubbing methods can be considered "mature" technologies. For gas volumes above approximately 50,000 acfm, the use of wet scrubbing techniques can become prohibitively expensive. Current techniques for high gas volumes primarily involve the use of electrostatic precipitators (wet or dry) and baghouses. These devices can be used to perform one function very well, the removal of particulate.

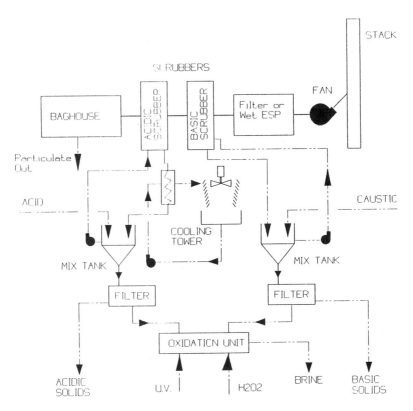

FIGURE 3.34. Hybrid scrubbing system.

For maximum particulate capture, a fabric filter collector or electrostatic precipitator (ESP) would be used in a conventional manner. It would *not* be used to remove both particulate and acid gases, for doing so inordinately increases the waste volume produced by the system. To reduce waste volumes, if a baghouse were used, the bag material would be selected for maximum particulate removal with a minimum need for precoat.

The particulate would be removed in this stage with little or no effort made to remove acid gases or volatile heavy metals. The collector would be sized for maximum particulate removal performance at operating temperatures compatible with the bag material. For example, a baghouse would operate well above the acid dewpoint of the gas stream and well below the maximum service temperature of the bag material.

If an ESP were used, it too would *only* be required to remove particulate and would be so designed.

3.11.1.1.2 Acid Gases

To remove acidic gases at high efficiency with minimal reagent use, wet scrubbing is simply without equal. Removal efficiencies of hydrochloric acid (HCl) of well over 99.9% have been recorded at a number of medical waste incinerator installations using wet scrubbing techniques. It is also widely known that near stoichiometric quantities of alkali can be used to remove these acidic components.

Wet scrubbing techniques (absorption) would therefore be suggested as the acid gas removal technique. High volume mass transfer devices such as spray towers or fluidized bed scrubbers [9] (Catenary Grid Scurbber) could be used to allow high removal efficiencies while producing low quantities of waste products. A soluble alkali such as sodium or magnesium hydroxide would be used to chemically react with the absorbed acids in conventional manner.

Figure 3.35 shows a fluidized bed type high velocity absorber module sized for 80,000 acfm at 3 in. w.c. pressure drop and 99 + % HCl capture. These high velocity absorbers can be designed to remove reactive acidic gases in both cocurrent and countercurrent gas contact modes. Multizone spray towers are being used successfully in this service in flue gas desulfurization systems worldwide and could serve the function of acidic gas absorption.

3.11.1.1.3 Heavy Metals

When volatile heavy metals capture is considered, opportunities are created to integrate the absorption equipment with heavy metals capture.

It is theorized that some heavy metals emissions are condensed upon flue gas particulate prior to primary particulate capture and can thus be removed along with the fly ash. As designers of combustion equipment labor to combust better, *less* residual particulate is created upon which the metals may condense thereby reducing the probability of capture.

Rather than rely upon this condensation or supplementing the target nuclei with injected sorbents that increase the waste burden, heavy metals capture can be improved upon by designing the collector to suit the physical parameters of the pollutant.

If our source emission contains mercury, for example, we would need to reduce the flue gas temperature and condense the mercury. Though some mercury would be condensed on the particulate in the flue gas, additional mercury would need to be removed. After condensation, we would like to react the captured mercury and remove it in its most stable form. Flue gas condensation scrubbing techniques have been widely applied for mercury

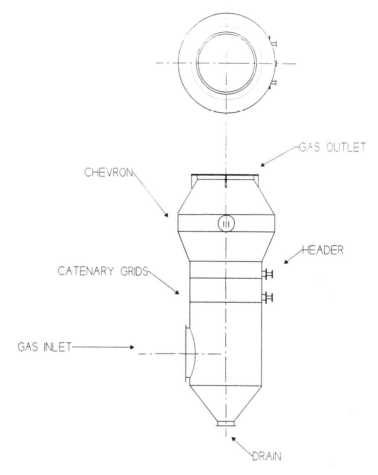

FIGURE 3.35. Fluidized bed scrubber.

on hot gas cleaning systems for hazardous waste incinerators both here and in Europe. Indeed, one of the oldest licensed hazardous waste incinerators in the United States (Rollins, Deer Park, TX) uses this technique [10].

Table 3.18 shows the relative volatility of various mercury compounds and their solubility. It is clear that converting the mercury to a sulfide is suggested rather than converting it to a chloride. The former is a much more stable salt than the latter. The condenser/absorber circuit can provide dual service by being operated at a low pH condition that yields a sulfide mercury residual.

Figure 3.36 shows the reduction in temperature of a hot gas cleaning system if condensation scrubbing techniques are utilized. Even the heat of compression of the fan is harnessed to help reheat the flue gas. Most importantly, the gas temperature is kept low at those capture zones wherein the contaminant (in this case mercury) is volatile.

Our Hybrid Scrubbing System would accomplish this by using a condensing absorber system using sulfuric acid as the scrubbing media at a pH of near 2.

In the condenser/absorber stage, the gases leaving the particulate collector are adiabatically saturated and subcooled using a low pH scrubbing liquid. The chemistry is designed to remove the mercury and to convert it to the more stable sulfite or sulfate form. After cooling, subsequent liquid/solids separations are kept suitably cool to reduce the volatilization of the captured mercury. If the source is very high in mercury, an irrigated fiberbed coalescing filter [11] or wet ESP may be needed to remove the condensed mercury particulate residual leaving the condenser/absorber.

Fiberbed technology has advanced significantly in recent years with the development of the nested prefilter arrangement. Particularly useful where high concentrations of toxic heavy metals exist, this technology could find application as a final security filter. Both the wet ESP and the fiberbed remove the residual pollutants mechanically, without additives.

Since the acidic scrubbing solution reduces HCl capture, a second absorber operating in a higher pH mode (pH 5.5 to 6.5) follows the first absorber and precedes the final collector.

The wastewater from the acidic heavy metals scrubber is sent to a physical separations unit which filters the captured sulfide from the liquid stream. The residual liquid is returned to the acidid scrubbing system loop.

TABLE 3.18. Properties of Mercury Compounds.

Compound	Decomposes @ (°C/F)	Melts @ (°C/F)	Boils @ (°C/F)
Mercuric:			
Carbonate			
Chloride		277/530	304/580
Oxide	d. 500/932		
Sulfate			
Sulfide	584/1082		
Mercurous:			
Carbonate	130/266		
Chloride		302/578	384/722
Oxide	100/212		
Sulfate			
Sulfide			

Source: *Perry's Chemical Engineering Handbook of Chemistry and Physics*, 45th Edition.

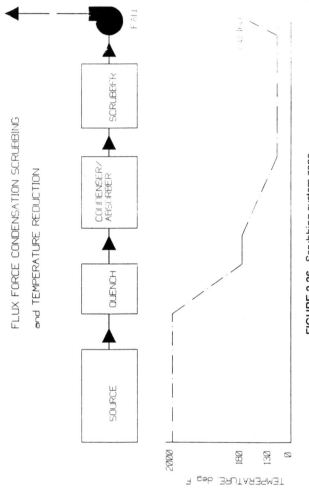

FIGURE 3.36. Scrubbing system zone.

The wastewater from the basic acid gas absorber is sent to its own wastewater treatment unit for separation. This unit is last in the gas collecting stream. A soluble alkali would be used (such as sodium hydroxide) and the waste product would be sodium chloride brine.

3.11.1.1.4 Toxic Organics

The waste streams from incinerators are known to contain small quantities of toxic organics. Some of these compounds, such as dioxins and dibenzofurans are treated as if they were potent human carcinogens though the supporting data for this reputation is conflicting or insufficient upon which to draw a definitive conclusion.

In keeping with the goal of our item (6), we would like to *destroy* these compounds, at the site, once captured. The literature also shows that these compounds can be removed through condensation techniques [12]. Tests on medical waste incinerators operating at low temperatures exhibited better toxic organics control than "dry" scrubbers operating at higher temperatures [13].

In recent years, wastewater treatment techniques have been developed that destroy toxic organics through the use of hydrogen peroxide catalyzed by ultraviolet light [14] (either continuous or pulsed). The wastewater from the condensing system would therefore be treated by this technique for on-site destruction of the captured toxic organics. The use of the hydrogen peroxide does not create additional residual solids which must be disposed of.

Looking at Table 3.19, we can compare our alternate system to that of the popular spray dryer and baghouse "additive" type system shown in Table 3.17. While the dry scrubber produces about 2.25 kg of "waste" per kg of contaminant, the hybrid scrubbing system produces only 1.77 kg/kg. It is clear that the technology currently exists to control these emissions more effectively while producing significantly less material which must be disposed of and producing a more stable residual for the more toxic substances.

Granted, this system is not as simple as injecting additives and throwing them away. It is far more complex and requires additional effort.

The question remains, however: "What is the *right* way to control these pollutants?" Use additives that *increase* the waste load to the landfill and which may be emitted later? Or make the extra effort to produce a minimal volume, stabilized waste product?

Will future generations have the need to ask, "Didn't they know what they were doing was wrong?" With a properly designed hybrid wet scrubbing system, they shouldn't have to ask.

TABLE 3.19. Wet Scrubber with Additives, Waste Produced.

Contaminant	Emission Rate, kg/hr	Waste Ratio, kg/kg	Additives/Waste			Total Waste Produced
			NaOH/NaCl	Sulfide	Peroxide	
Particulate	388	0	0	0	0	388
HCl	854	1.63	1392	0	0	1392
SO$_2$	2026	1.97	0	3991	0	3991
HgO	2.3	1.29	0	3	0	3
Dioxins	0.001	3	0	0	0.003	0.003
Minimum 2aste = 3270 kg/hr						5774
Ratio of total waste produced to minimum waste produced =						1.77

Basis: 170,000 Nm^3/hr gas flow rate
2000 ppmv SO$_2$, SR = 1.0
1500 ppmv HCl, SR = 1.0
0.005 mg/Nm^3 HgO
40 ng/Nm^3 dioxins
Assume 100% capture of contaminants.

3.12 REFERENCES

1. "Environmental Pollution Control at Hot Mix Asphalt Plants" (Riverdale, MD: National Asphalt Pavement Association), Information Series No. 27.

2. McIlvaine, R., Ed. *The McIlvaine Scrubber Manual* (Northbrook, IL: The McIlvaine Company, revised monthly).

3. Hohlfeld, R. W. "SO_2 Emission Reduction Caustic Scrub," *Inorganic Chemicals* (Walnut Creek, CA: Dow Chemical, USA, 1987).

4. Moller, J. T., Christiansen, O. B., "Dry Scrubbing of Hazardous Waste Incinerator Flue Gas by Spray Dryer Absorption," AWMA, 77th Annual Meeting, Paper 84.9.5, June 24–29, 1984.

5. Rood, M. J., Wood, J. P., "Modeling the Simultaneous Removal of SO_2 and HCl from Municipal Waste Combustion Flue Gas via Spray Dryer Absorption," AWMA Paper 92-39.02, 85th Annual Conference, June 1992, Figure 8.

6. Hartenstein, H.-U. "A Fixed Bed Activated Coke/Carbon Filter as a Final Gas Cleaning Stage Retrofitted for a Hazardous Waste Incineration Plant—The First Six Months' Operating Experience," AWMA Paper 92-48.04, Annual Meeting, June 1992.

7. Quimby, J. M., Teller, A. J. "Mercury Removal from Incinerator Flue Gas," AWMA Paper 91-35.5, 84th Annual Meeting, June, 1991.

8. Lemann, Dr. M., "The Possibilities of Adding a Special Wet Scrubber System for MSW Facilities," Von Roll, AG, Zurich, Switzerland.

9. Hesketh, H., Schifftner, K., "Describing and Defining the Catenary Grid Fluidized Bed Mass Transfer Separator," *Atmospheric Environment,* Vol. 19, No. 10, Pergamon Press, 1985, pp. 1565–1572.

10. Calvert, S., Ed., "Feasibility of Flux Force/Condensation Scrubbing on Fine Particulate Collection," U.S. EPA Technology Series, EPA G50/2/73-036, Oct. 1973.

11. Taub, S., Fowler, E., "The Role of Fiber Bed Filters in Waste Minimization," CECO Filters, Inc., Conshohocken, PA, Technical Publication 103/90, 1990.

12. Schifftner, K., "Integrating Incinerator, Scrubber and Waste Disposal: The Key to Successful Medical Waste Control," AWMA Paper 91-30.2, 84th Annual Meeting, June 1991.

13. California Air Resources Board, "Final Draft—State of the Art Assessment of Medical Waste Thermal Treatment," Energy and Environmental Research Corporation, Irvine, CA, Contract A832-155, Dec. 1989, pp. 150.

14. Personal Communication, Mr. Emory Froelich, Peroxidation Systems, Chandler, AZ, Nov. 1992.

CHAPTER 4

Maintenance and Control

4.1 INTRODUCTION

Few systems of any type run unattended and without maintenance. Though operator presence is not required in many scrubbing installations, the operator can be there, "in spirit" at least, through the use of instrumentation and the implementation of preventive maintenance procedures.

This short chapter is provided to suggest basic procedures that should be considered relative to industrial scrubbers and scrubbing systems. Manufacturers' literature and other references [1,2] deal extensively with this subject.

4.2 OPTIMIZATION

There are obvious reasons as to why the gases are being scrubbed and to what level the cleaning must extend. Beyond that, however, there are many items that may make it cheaper to achieve the same degree of control but at a reduced cost. The most significant economic penalty is gas pumping costs, which are a direct function of the gas phase pressure drop. This relates to the type of scrubber, operating conditions, efficiency and the liquid rate—so all portions of the system are involved.

In planning a scrubbing system there are many direct and indirect factors to be considered. Some of these are listed in the sections below:

(1) Characterize inlet flue gas stream regarding the following:
   • particle size
   • particle size distribution

- particle quantity
- gas temperature
- gas quantity
- gas composition (especially chlorides)
- dew point (water, acid, other)

Attempt to simplify the scrubber by use of a more open unit.

(2) Evaluate mist eliminators:
- Horizontal gas flow devices have more capacity but require more room.
- Vertical gas flow devices need sharp angles to help drainage and are not as effective as the horizontal flow types.

(3) Material of construction:
- 316L or 904L or others are needed if chlorides are present.
- Plastic and FRP linings may have lower coefficients of friction and less weight.

(4) Gas ducts:
- Use a gas velocity adequate to prevent particles from settling but low enough to minimize abrasion (e.g., ~50 ft/sec).
- Use flexible expansion joints.
- Provide drains for all lines.
- Slope lines to drains.
- Clean out access doors.

(5) Fan:
- Use replaceable blade liners.
- Wash provisions for cleaning blade tips.
- Make sure drains are provided and functional [also see item (14) below].

4.2.1 Liquid Circuits (Use Figures 4.1a and 4.1b as a Guide)

(1) Feed solids-carrying headers from the *bottom* and discharge from the *bottom*. Place these headers above the functional section of the scrubber (venturi, spray nozzles, etc.), permitting them to drain back upon shutdown.

(2) Provide individual shutoff valves for each spray nozzle (or group of nozzles removed as a section) with a pressure gauge or flow indicator to monitor flow. This simplifies diagnosing nozzle problems and performing service.

(3) Avoid internal sumps, weirs, cavities or baffles if at all possible. These devices are typically subject to double-sided corrosion, wear and unseen failure.

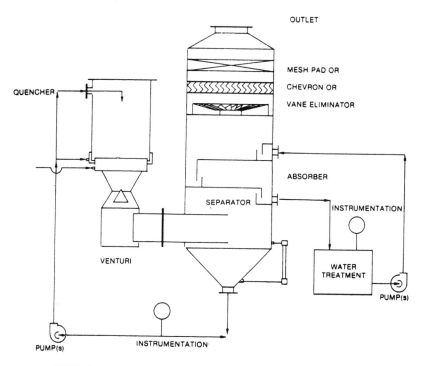

FIGURE 4.1a. Scrubbing system liquid circuits—basic arrangement.

(4) Provide adequate drains. If solids are present, provide a rod-out connection on the scrubber drain, avoiding right angle bends. A 45° lateral provides a good rod-out connection. Always drain a solids handling scrubber if a prolonged shutdown is expected.

(5) Avoid the use of spray nozzles if possible. They plug and wear. If you must, use plugging-resistant types (such as Bete Fog nozzles).

(6) Do not place liquid or level instrumentation in turbulent zones. Shield them with a baffle or a vented, turned down elbow in the vessel.

(7) Carefully investigate the drain system. Avoid placing scrubber discharge into drains, which may cause a chemical reaction and/or pollution problems. Many times, a dedicated drain is needed to direct the flow to a proper place of treatment (neutralization tank, clarifier, etc.).

(8) Allow for release of scrubber holdup of the liquid. High suction scrubbing systems sometimes lift the water level in the scrubber sump zone, causing an excessive volume upon shutdown. This is especially

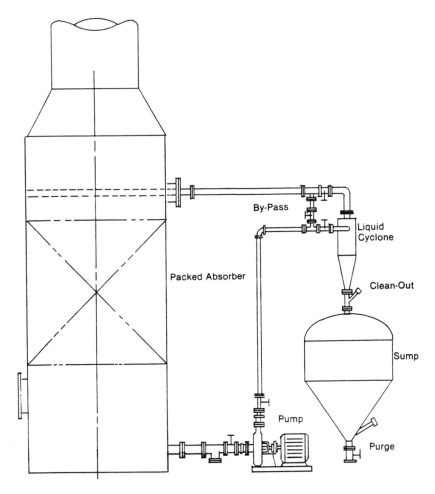

FIGURE 4.1b. Scrubbing system liquid circuits—modification.

true where an open drain pipe goes to a liquid/solids settling tank, for example. The lowest point must have enough free board to accommodate a sudden discharge of scrubbing liquid.

(9) Provide water hammer preventers on plastic spray nozzle lines where used and on intermittent duty lines (fan sprays, for example). Water hammer has broken many nozzle assemblies and headers over the years.

(10) Make all nozzle assemblies retractable (bayonet-type fittings) for simplified maintenance.

(11) Use gauge guards or snubbers on all pressure gauge connections, especially when handling solids. Place a tee below the snubber with a valve connected to a source of fresh water at higher pressure than the header. This provides a blow-back which can clear the lines.

(12) Use reliable check valves on all chemical feed lines with bypasses and shutoff valves at points of service. Support connections to pumps by connecting to the same bracket system which supports the pipes (not the pump). Smaller chemical feed lines tend to vibrate at pump operating speeds. The additional mass of the piping support steel usually provides adequate dampening.

(13) Use expansion joints at all pump inlets and discharges, if possible.

(14) Seal all fan or stack drains to adequate barometric legs.

(15) Remember to provide a large fill line for the scrubber.

(16) Level controls should be accessible. Avoid using flat type on sticky substance applications. Conductance or capacitance types generally work best, but must use a stilling basin in most scrubbers.

(17) Provide sample draw-offs where required.

(18) Flange sections of pipe which may fill with solids or readily corrode or abrade (this facilitates removal). Some systems use Victaulic-type couplings.

(19) Mark all caustic, acid, or hot water lines to protect your personnel.

(20) Do not place pipe sections in difficult-to-reach areas. Use unions or couplings to facilitate disassembly and inspection.

(21) Maintain adequate velocities in lines to prevent particle settling but keep velocities low enough to minimize pipe wear (e.g., 4–7 ft/sec).

(22) Consider rubber-lined or plastic pipes to reduce friction loss and wear.

(23) Slope liquid lines to drains.

(24) Heat trace lines if short shutdowns in very cold weather may occur and lines are not drained.

4.3 SYSTEM CHARACTERISTICS

It may be useful to provide a comparison of the various types of scrubbing systems. The following listing summarizes some of these:

- *Plate tower*—can easily put clean-out door on every plate if desired. The disadvantages are rapid scale formation if the liquid is not adequate, and the plate bottom may require washing.

- *Packed tower*—not for use with particulates (any solids must be washed out to prevent plugging). Ceramic packing cracks at rapid temperature change and is heavy plastic. It is temperature-limited.
- *Ejector*—high-velocity system with rapid wear.
- *Spray tower*—nozzle wear can be rapid. Ceramic may be used to prevent this, but it may crack during tightening (use a plastic/Teflon® thread cover to prevent this).
- *Mist eliminator*—most often use *continuous* under and over sprays.
- *Venturi* (gas atomized)—use silicon carbide brick lining for high velocity and ΔP applications.
- *Impingement and entrainment*—trouble keeping liquid levels at proper height; they are heavy.
- *Mechanically aided*—the mechanical impeller used to splash the liquid scales easily and requires frequent maintenance.
- *Moving beds*—marble bed and ping pong ball type can get local buildups and high-velocity results. Plastic balls wear and marbles crack and fall through.
- *Foam*—is very troublesome when foam destructor fails.
- *Steam-assisted*—heat exchanger on inlet plugs easily.
- *Charged*—has the problems common to both ESPs and scrubbers plus corrosion due to charged liquids.
- *Fan*—not usually a good application and may require weekly cleaning.
- *Prequench*—has many advantages besides conditioning, e.g., foundry cupola with no prequench needs a new scrubber every 6 months while the next cupola with quench operates at 100% for over 4 years.

4.4 CHEMICAL CONTROL

Wet scrubbers are basically chemical process equipment. As such, a good chemical and chemical engineering background can be very helpful to provide successful operation. Some systems, especially the FGD, odor control and sludge scrubbers, have definite chemical requirements and control. For example, lime scrubbers will scale up above and below the pH limits of about 7–9. For limestone, the pH limits are about 5.5–6. In such systems the pH is lowest in the scrubber just upstream of the holding tank. The pH is highest just as it enters the scrubber and changes as reactions occur. In many systems, time is required for reactions to occur and go to completion, so pH also varies with time.

Excess chemical achieve increased costs by: reducing the utilization of the reagent; possibly forming scale; absorbing unnecessary gases; removing excess pollutant; and possibly adversely affecting the material of construction. Insufficient chemicals cause some of the same problems and also inefficient removal.

It is difficult to obtain good measurements of chemical concentrations in a scrubber. As noted, pH is a preferred technique. However, the sensor must be frequently removed, cleaned and calibrated. Some systems require flow-through sensing arrangements for accurate readings. This results in high wear and may necessitate frequent line cleaning.

4.5 OTHER INSTRUMENTATION

The importance of pressure drop (ΔP) in establishing scrubbing efficiency has been discussed. ΔP is also important as the clue to how the unit is performing. Thus the most important instruments are ΔP and static pressure instruments. These must be on the scrubber, mist eliminator and fan as a minimum. The connections should be up, out of the liquid, and designed for minimal closure by scale blockage. A suggestion for this is given in Figure 4.2. The pressure lines should also have clean dry air purge provisions to keep the lines and openings free. These may be continuous or intermittent using 3-way valves to protect the instruments.

4.6 PERSONNEL

Scrubbing systems, as chemical processing equipment, do not require large numbers of operation/maintenance personnel. However, these individuals should be trained and capable of performing the required functions.

As systems become very large, the number of operators and maintenance persons does not increase in relation to size. For example, the huge Sherburne FGD system, which is 1400 MW, requires an average of only 3 persons per each 100 MW scrubbing capacity per shift. This includes 16 operators, 8 maintenance persons, 4 instrument technicians, 2 chemists, 2 engineers, 1 electrician and 22 laborers. Similar skills are required for small systems, and more personnel are needed per unit of system capacity.

4.7 CHECKLISTS

Many checks can and should be made on scrubbing systems. Table 4.1 provides a suggested maintenance checklist with frequencies as noted. This

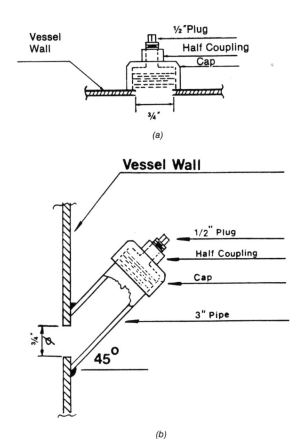

FIGURE 4.2. Suggested pressure tap connection arrangement for wet scrubbing systems: (a) horizontal system, (b) vertical system.

182

TABLE 4.1. *Suggested Exhaust Scrubber Maintenance Checklist.*

Daily Inspection	
Check Point	What to Look for
1. Pump	a. Leaking at gland
	b. Increased noise
2. Valves	a. Position
	b. Leaks
3. Piping	a. Leaks
4. Body	a. Leaks
5. Pressure gauge	a. Pressure change from previous day
6. Pressure gauge ammeter combination	a. Changes in either or both pressure reading and ampere draw from last clean system check readings
Gauge Readings	Probable Trouble
1. a. Water pressure same	
b. Ampere draw same	a. None
2. a. Water pressure decrease	a. Missing nozzles
b. Ampere draw decrease	b. Pump wear or plugging of suction line
3. a. Water pressure same or increase	
b. Ampere draw decrease	a. Plugging of nozzles or spray bars
4. a. Water pressure increase	
b. Ampere draw increase	a. Holes in spray bar or manifold
Weekly Inspection	
Check Point	What to Look for
1. Spray bars	a. Plugged nozzles
	b. Worn or missing nozzles
2. Pipes and manifolds	a. Plugging or leaks
3. Pressure gauge	a. Check accuracy
4. Pumps and valves	a. Wear
	b. Valve operation
5. Main body of scrubber	a. Material feed buildup
	b. Abrasion
	c. Corrosion

TABLE 4.2. Scrubbing System Checklist.

Valves/dampers
    bypass (periodically operate to be
        sure they are not frozen in
        open or closed position)
    density control
    water purge control
    pH elements, flush water
    chemical makeup control
    pond return
    fresh water makeup
 	
Process water
    pumps
    pressure gauges
    tanks
    agitators
 	
Controls/feedback
    stock gas flow
    makeup water
    reagent feeder rate
    slurry pH
    tank level
    slurry density
    by-product disposal
	Safety systems
	    high-temperature limit controls
	    interlocks
	    alarms
	
	Compressed air
	    availability
	    filters
	
	Sensors
	    pressure
	    temperature
	    flow
	
	Heat tracing
	
	Analytical systems

could be tailored to fit specific situations. In addition to this, the list of items to check in Table 4.2 may be useful. Furthermore, files should be available for important items such as:

- operating manuals
- system and component drawings
- parts lists
- schedules
- manufacturers's specifications

4.8 REFERENCES

1. Cross, F. L. and H. E. Hesketh. *Handbook for the Operation and Maintenance of Air Pollution Control Equipment* (Lancaster, PA: Technomic Publishing Co., Inc., 1975).

2. Young, R. A. and F. L. Cross. *Operation & Maintenance for Air Particulate Control Equipment* (Ann Arbor, MI: Ann Arbor Science Publishers Inc., 1980).

CHAPTER 5

Testing Wet Scrubbers

5.1 INTRODUCTION

As with all types of air pollution control devices, a knowledge of wet scrubber inlet and outlet data is necessary to maintain a "healthy" operation and to troubleshoot a "sick" system. Accurate outlet wet scrubber data are more difficult to obtain if you do not follow certain procedures. For example, mist eliminators must be properly secured in place, operated with correct gas velocities, and otherwise made to function as designed so as not to throw excess moisture droplets into the scrubber outlet gas stream. All water droplets will be accounted for as water vapor and particulate matter. This could severely alter the test results, as well as cause the emissions to exceed the allowable limits.

The test gas from a scrubber must be kept above the moisture dew point to protect the filter paper. Moisture, whether it be from collected or condensed droplets, can ruin the test run by causing the filter paper to fail.

Pressure drop measurements across both the scrubber and the mist eliminator have already been noted as necessary to the proper operation of the scrubber. This also applies when testing in that they are a necessary part of the test data.

5.2 DRY FLUE GAS COMPOSITION

For most typical flue gases, the dry gas analysis consists of CO_2, O_2 and N_2. This includes most combustion gases and many process gases. However, if there is a "significant" amount of CO, SO_2 or other gas, then they must be accounted for. Normally, anything less than 1000 ppm (i.e., 0.1%) by volume is negligible for the calculation of molecular weight, which is a main purpose for dry gas analysis. Nitrogen concentration is obtained by difference from 100%. Don't forget, water vapor must be included to obtain the wet molecular weight, which is used in the pitot equation. This is explained in Section 5.3.

The dry gas analysis can be obtained by an Orsat type analysis. Currently, the more commonly used system is the Bacharach Gas Analyzers, which go by the trademark name of Fyrite. These units are rugged, relatively leak-proof, "dumbbell" shaped absorbers. They are filled with absorbing solution available from Bacharach, Inc. (Pittsburgh, PA). They come in a carrying case with a gas pumping bulb, filter and tubing for normal field use. Fyrite operating temperatures are from -30–$150°F$ and gases up to $850°F$ can be sampled. They can be obtained in scale ranges of 0–7.6%, 0–20%, and 0–60% CO_2 and O_2.

Operation of Fyrite absorbers is summarized in Table 5.1. The O_2 absorber can be checked by testing ambient air, which contains 20.9% O_2. The CO_2 absorber can be checked using human breath, which should consist of about 4% CO_2. Other operational notes for Fyrites are:

- *Fluid level* – 1/8 to 5/8 in. of absorbing liquid in bore when zeroed. Raise level by dripping clean water onto top and working plunger. Lower level by unscrewing top and removing with syringe or glass rod.
- *Temperature* – allow 15–20 minutes to equalize before starting tests, as long as gas is $<850°F$; sampling assembly permits adequate cooling between successive tests, if sample intervals are >5 min.
- *Sample lines* – if lines on system provided are increased, add one bulb squeeze for each 3 in.³ of added line.

One word of caution regarding dry flue gas and wet scrubbers is that wet scrubbers frequently operate with alkali slurry. This absorbs CO_2, therefore a carbon balance across inlet-outlet conditions cannot be made accurately. Obviously, a water balance cannot be made directly. This means that an oxygen balance may be the only option.

5.3 DIAGNOSTIC TESTING

This type of testing can help determine whether a wet scrubbing system is operating at optimum conditions. The following data are useful for this:

- scrubber size and configuration

TABLE 5.1. Condensed Operating Instructions for Fyrite Test Indicators.

CO_2	O_2

Zero Scale
  Depress plunger one to two
    times
  Full rotation one time
  Hold @ 45° 5 sec, then upright
  Depress plunger
  Set scale Zero at liquid top
	Zero Scale
	  Depress plunger one to two
	    times
	  Full rotation two times
	  Hold @ 45° 5 sec, then upright
	  Depress plunger (if fluid drops
	    over 4%, repeat two rota-
	    tions, then depress plunger
	    again)
	  Set scale Zero at liquid top

Obtain Pumped Sample
  Lay connector on top of each dumbbell in turn and push to depress
    plunger.
  Squeeze bulb 18 times, then remove connector *during* last bulb squeeze.	

Absorb CO_2
  Full rotation two times
  Hold @ 45° 5 sec, then up
  2 sec
  Read CO_2 on scale
	Absorb O_2
	  Full rotation four times
	  Hold @ 45° 5 sec, then up
	  2 sec
	  Read O_2 on scale

Finish
  Depress plunger
  Check-re zero
	Finish
	  Depress plunger (if O_2 was
	    >4%, rotate four times,
	    then depress plunger again)
	  Check-re zero

Fluid Strength Check
  Don't depress plunger, but
    repeat "absorb CO_2," it is
    OK if change <1/2% CO_2
	Fluid Strength Check
	  Same as CO_2

- wet and dry bulb temperatures
- Orsat (or Fyrite) dry gas analysis
- pitot pressure drop
- static pressure
- barometric pressure
- spray nozzle header pressure
- liquid flow rates (recycle, bleed, makeup)
- liquid solids concentration
- alkali to acid stoichiometry
- liquid pH
- spray nozzle type(s) and number

The first six items of the list above also produce the necessary data to obtain gas flow rates. This plus the liquid data yield liquid-to-gas ratio (typically expressed as gal liquid per 1000 actual ft³ of gas).

One of the first steps in checking a "sick" wet scrubber is to see if the scrubber exit gas is saturated. This must be done before any fans and as near the scrubber exit as possible. If the mist eliminators are working properly and the wet and dry bulb temperatures are the same, the exit gas is probably saturated. If the gas is not saturated, the scrubber is not working or is improperly designed. If the scrubber cannot even saturate the gas, then it cannot do a good job of absorbing gases or removing particles.

Steam tables and psychrometric charts (see Section 1.3.2) are good materials to have available on a test. If the gas is saturated, the water content can be found by looking up the saturation vapor pressure of water in the steam tables at that temperature and dividing by the absolute pressure (this is barometric pressure ± the duct pressure at the test location). If the gases are not saturated, air psychrometric charts can be used to find humid content, H, expressed as lb water per lb dry air. To do this, start at the dry bulb temperature and follow the adiabatic cooling curve until the wet bulb temperature is reached. This is H. Find % water content by

$$\% \text{ water} = \frac{161H}{1.61H + 1}$$

The pitot equation is used to find the gas velocity. The dry gas analysis plus the water content are used to calculate the wet gas molecular weight, MW_{wet}. This pitot equation is

$$V_s = 5128.8C_p \sqrt{\frac{(460 + T_s)(\Delta P)}{P_s(MW_{wet})}}$$

where

V_s = actual gas velocity in ft/min
C_p = pitot coefficient dimensionless
    (= 0.99 for std. pitot and 0.84 for properly constructed S pitot)
T_s = stack temperature, °F
ΔP = pitot pressure differential, inches of water
P_s = stack or duct pressure, inches of mercury
    (this is barometric pressure ± duct pressure)

Gas velocity times the sample duct cross section in ft² gives gas actual flow rate in cubic feet per minutes (acfm). Considering the gas to be an ideal gas, the standard cfm (scfm) can be calculated remembering that scfm is given at 68°F and 1 atm:

$$scfm = (acfm)\frac{460 + 68}{460 + T_s}\left(\frac{29.92}{P_s}\right)$$

The dry scfm (dscfm) can be obtained by

$$dscfm = (scfm)\left(\frac{100 - \%H_2O}{100}\right)$$

It is dscfm that is most useful to check scrubber behavior. For example, if the system has no leaks, dscfm in essentially equals dscfm out for typical systems. If dry gas MW differs significantly, it must be included in the inlet-outlet balance.

Pressure gauges on the headers of spray bars can be used to find the liquid gpm in a scrubber when the number and type of nozzles are known. This information can be found in the manufacturer's literature. Flow meters can also provide these data, but may be difficult to read and keep in operation, because recycle liquids usually contain 10% or more solids by weight in the slurry.

Liquid-to-gas (L/G) ratios can then be calculated from these data. As a rule of thumb: for particle scrubbing, the L/G should be no lower than 5 gal/1000 acf and for gas absorption, L/G should be about 25 or greater.

Also, check liquid rates, nozzle types and numbers and scrubber configuration to be certain that the liquid in the scrubber is spread across the entire scrubber cross section area, if required. The manufacturer's literature on the spray nozzles can help in this estimation.

Solids are often necessary in the recycle system. Some chemicals are slow to react after being absorbed, and it may take precipitates 10–20 minutes to form. The solids in the slurry can be used for nucleation sites for precipitates to form, thus saving the scrubber from having precipitates adhering to the scrubber internals, which results in plugging of the scrubber.

Frequently, scrubbing systems are used to remove acid components from the inlet gases. The absorbed acid gases must be neutralized by addition of an alkali (e.g., lime). The balanced stoichiometric chemical equation will show the moles of alkali needed to remove one mole of acid. This is a stoichiometry of 1. Sometimes it is desirable to use a stoichiometry of about 1.2 to assure adequate neutralization. However, an excessively high stoichiometry would be a disadvantage as chemicals are often the most expensive item in wet scrubber operation.

Chemical stoichiometry can be used to estimate amounts of alkali required, but pH is often used to control the actual chemical addition, because flows and concentrations may fluctuate. Based on the remarks in the previous paragraph, good pH control can be very important to assure adequate acid removal and to keep costs down.

5.4 COMPLIANCE TESTING

Many components for compliance testing are described in Sections 5.2 and 5.3. Compliance testing procedures are described in numerous publications such as Hesketh [1], in the *Federal Register* [2] and in state approved testing procedures. The official approved procedure(s) should be used unless specific approval is obtained before the test for any variations in test procedures.

Basically, EPA Method 5 is used for particulate sampling. This includes: Method 1 – traverses; Method 2 – gas velocity; Method 3 – CO_2, O_2, excess air and MW; and Method 4 – moisture. In addition, any of the following may be needed:

- Method 6 SO_2
- Method 7 NO_x
- Method 8 H_2SO_4 and SO_2
- Method 9   opacity
- Method 10   CO
- Method 11 H_2S
- Method 19 SO_2 and other gases from utilities

Other methods are also available.

The Method 5 train is the work horse of testing. The system can be modified to measure organics (VOST or MM-5) and HAPs in addition. Figure 5.1 is a schematic of the Method 5 train (M-5). The M-5 train must be assembled leak tight. Generally, it is not permitted to leak more than 0.02 ft³ in one minute at 15 in. Hg vacuum when checked at the filter inlet. It can leak no more than 0.02 cfm at 5 in. Hg when checked at the end of the test nozzle. The dry gas meter (DGM) is leak checked and calibrated at 21 in. Hg vacuum before going to the field for use and the DGM calibration factor is secured to the M-5 control box.

The M-5 hot box must be kept at 220–270°F during the test to help assure that the filter is kept dry and that the catch is not overheated and volatilized. Impingers in the cold box are submersed into an ice water bath to keep them as cool as possible. Normally, the S-type pitot is constructed to the 0.84 pitot coefficient.

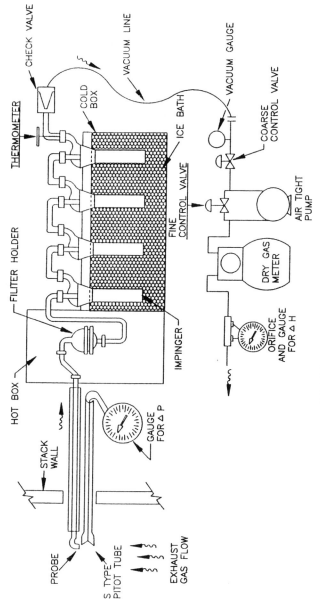

FIGURE 5.1. U.S. EPA Method 5 test train.

191

When testing wet scrubber outlets, make certain that all lines (e.g., sample lines, pitot lines, etc.) are kept clear of condensed and/or thrown water. The pitot lines are especially vulnerable to this and should be blown out periodically as necessary.

5.5 REFERENCES

1. Hesketh, Howard E., *Air and Waste Management—A Laboratory/Field Handbook,* Lancaster, PA: Technomic Publishing Co., Inc. (1994).
2. *Code of Federal Regulations,* Title 40—Protection of the Environment, Chapter 1—EPA, Subchapter C—Air Programs, Part 60—Standards of Performance for New Stationary Sources, Appendix A—Reference Methods (as amended).

CHAPTER 6

Suggested Readings

Although applications engineers must use a wide variety of reference material in order to select the optimum technology to solve a particular problem, there are a few resources that are repeatedly used.

Kept within reach or on a nearby shelf, these reference sources can be consulted to help solve the majority of air pollution control problems that the applications engineer might face. Some or all of these resources may be of value to you if you are called upon to research a pollution problem or to seek a solution to one.

The sources we use follow (a list of where to obtain this reference material is located at the end of this chapter):

(1) *Industrial Ventilation*

Published by the American Council of Governmental Industrial Hygienists, this book contains basic information regarding the suggested method of collection of contaminant vapors and particulate using proven techniques. Beginning with the basic physics of air movement and the techniques used for measuring air flow and pressure, it explains the sizing of ventilation systems, discusses fan application and design, details the operation of basic air pollution control equipment, and outlines many typical ventilation system applications.

(2) *Fan Engineering*

This book is published by the Buffalo Forge Company, makers of industrial fans and blowers. The book is an excellent reference source for the basics of gas humidification and cooling, stream blending, fan selection and design, gas cleaning equipment sizing and operation, and equipment selection. The book has been used by air pollution control engineers for over four decades and can be seen on a number of consulting engineers' book shelves.

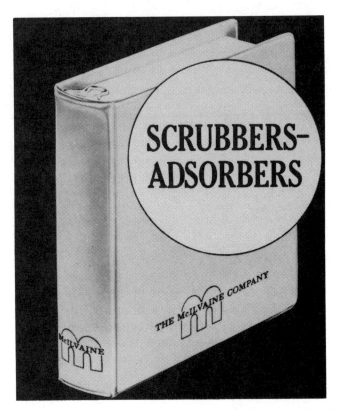

FIGURE 6.1. Courtesy of the McIlvaine Company, 2970 Maria Avenue, Northbrook, IL 60062 [Tel: (708) 272-0010].

(3) *McIlvaine Scrubber Manual*

The *McIlvaine Scrubber Manual* (Figure 6.1) is a multi-binder reference source on air pollution and its control. It is a subscription service that provides industry and marketing information in the form of newsletters, audio- and videotapes, and other formats. Found at most scrubbing system suppliers, many "end-users" also subscribe to the *McIlvaine Scrubber Newsletter* in order to keep current on technology and application trends.

(4) *Air Pollution Engineering Manual*

First published by the government as *AP-42, Air Pollution Engineering Manual,* this compendium of air pollution control application basics and case history type industry overviews was recently rewritten by a group of authors under the direction of the editors Mr. Anthony

Buonicore, a noted consultant, and Dr. Wayne Davis, a "student" of the industry, author, and professor. The book was compiled with the assistance of the Air and Waste Management Association (AWMA) and is published by Van Nostrand Reinhold. It is a valuable resource for those people trying to investigate the type of technology that has been used to date to solve a wide variety of air pollution problems. Many of the individual authors are noted experts in their areas of study.

(5) *Psychrometric Tables and Charts*

Also called the "Zimmerman and Lavine" book, this thin book contains the basic information needed to determine gas properties at various pressures and humidities. It contains psychrometric charts and temperatures and pressures beyond typical ambient conditions and has a simple and comprehensible description of how these conditions are calculated.

(6) *Cameron Hydraulic Data*

This book has been used for decades as a valuable source of information on piping, pumping, and hydraulic design.

(7) Various Corrosion Guides

Various pump, valve, elastomer, and resin manufacturers produce excellent guides that describe the corrosion resistance of various materials of construction used in air pollution control equipment. These guides are typically free for the asking from the vendors. A suggested method of using these is to acquire as many as you can and take a "straw poll" of the predicted corrosion resistance of the materials of construction that you might consider for an application. Not all of these tables agree, but "majority rule" can usually lead to the best selection for your application.

There are many other reference books of high quality that can also be used that are too numerous to mention. If you earn your living solving air pollution control problems, you will likely develop a short list of your own favorite reference sources and find yourself reaching for them in those all too frequent times of need.

(1) *Journal of the Air and Waste Management Association*
P.O. Box 2861
Pittsburgh, PA 15230

(2) *Pollution Engineering Magazine*
1350 E. Toughy Ave.
Des Plaines, IL 60017-5080

(3) *Power*
11 West 19th Street
New York, NY 10011

(4) *Environmental Science and Technology*
American Chemical Society
1155 16th Street NW
Washington, DC 20036

(5) *Technical Association of Pulp and Paper Industry (TAPPI)*
One Dunwoody Park
Atlanta, GA 30338

(6) Pollution Equipment News "Catalogue and Buyers Guide"
8650 Babcock Boulevard
Pittsburgh, PA 15237

(7) Bete Fog Nozzle Catalogue
50 Greenfield Street
Greenfield, MA 01302-0311

(8) Dwyer Instrument Catalogue
Dwyer Instrument Co.
Box 373
Michigan City, IN 46360

(9) Crane Co. Catalogue (Engineering Section)
Local Representative or
Crane Co.
300 Park Avenue
New York, NY 10022

(10) *Air Pollution Control—Traditional & Hazardous Pollutants, 2nd Ed.*
Howard Hesketh
Technomic Publishing Co., Inc.
851 New Holland Avenue
Box 3535
Lancaster, PA 17604

(11) Huntington Alloys Corrosion Chart (Nickel Alloys)
Local Representative or
Huntington Alloys, Inc.
Huntington, WV 25720

(12) A Guide to Corrosion Resistance
Climax Molybdenum Company
One Greenwich Plaza
Greenwich, CT 06830

(13) Derakane Chemical Resistance Table
Dow Chemical Company
Designed Products Dept.
2040 Willard H. Dow Center
Midland, MI 48640

(14) ATLAC Guide to Corrosion Control
Reichhold
P.O. Box 19129
Jacksonville, FL 32245

(15) Stainless Steel in Gas Scrubbers
Committee of Stainless Steel Producers
American Iron and Steel Institute
1000 16th Street NW
Washington, DC 20036

(16) *The McIlvaine Scrubber Manual*
The McIlvaine Co.
2970 Maria Avenue
Northbrook, IL 60062

(17) *Air Pollution Engineering Manual*
Anthony J. Buonicore and Wayne T. Davis, editors
Van Nostrand Reinhold
115 Fifth Avenue
New York, NY 10003

(18) *Fan Engineering—An Engineers Handbook on Fans
and Their Applications*
Buffalo Forge Company
465 Broadway
Buffalo, NY 14204

(19) *Handbook of Separation Techniques for Chemical Engineers*
Philip A. Schweitzer, editor
McGraw-Hill Book Company
1221 Avenue of the Americas
New York, NY 10020

(20) *Industrial Research Service's Psychrometric Tables and Charts*
O. T. Zimmerman, Ph.D., Irvin Lavine, Ph.D.
Industrial Research Service, Inc.
Dover, NH
(Mack Publishing, Easton, PA)

(21) *Industrial Ventilation — A Manual of Recommended Practice*
American Conference of Governmental Industrial Hygienists
6500 Glenway Avenue, Building D-7
Cincinnati, OH 45211

(22) *Cameron Hydraulic Data*
C. R. Westaway and A. W. Loomis
Ingersoll-Rand
942 Memorial Parkway
Building 101
Phillipsburg, NJ 08865

(continued)

Kenneth C. Schifftner is Technical Director, Compliance Systems Int. (Carlsbad, CA), a consulting firm specializing in air pollution control, energy conservation, problem solving, system design, and system upgrading. He received his degree in mechanical engineering from the New Jersey Institute of Technology in 1970.

Inventor of the Catenary Grid Scrubber® and NGV™ scrubber, Mr. Schifftner has been active in the field of air pollution control for nearly thirty years. He has written on a variety of topics, including odor control, lime sludge kiln scrubber operation, writing of specifications for air pollution control equipment, system operation and maintenance, and testing procedures.

Mr. Schifftner has held offices on committees in the American Society of Mechanical Engineers (ASME) and the Technical Association of the Pulp and Paper Industry. His research interests include further development and applications of the Catenary Grid Scrubber, the upgrade and repair of existing air pollution control systems, and the development of new systems for the microelectronics industry.

Howard E. Hesketh is Professor of Engineering (retired), Air Pollution Control, at Southern Illinois University at Carbondale and author of the first air pollution textbook (originally published by Ann Arbor Science Publishers). He is also a consultant and advisor to the U.S. EPA, U.S. Department of Commerce, and other agencies, universities, organizations and industries worldwide.

Dr. Hesketh received his Ph.D. in chemical engineering from the Pennsylvania State University, where he was a PHS Special Fellow. He is a Diplomate of the American Academy of Environmental Engineers and a registered Professional Engineer in Illinois and Pennsylvania. In addition to his practical and teaching experiences in air pollution, he received formal training in this area at the Pennsylvania State University Center for Environmental Studies.

Prior to his teaching, Dr. Hesketh was a Senior Chemical Engineer with DuPont, where he was co-inventor of a fluidization and conditioning process for crystalline particles. He also worked with The Beryllium Corporation and Bell Laboratories for Western Electric.

He is a member of several professional societies, is a past director and vice president, Air and Waste Management Association. He is a former Associate Editor for ASME Transactions Journals, and has co-edited, authored or been contributing editor to several environmental texts. Dr. Hesketh is author of *Fine Particles in Gaseous Media* and *Air Pollution Control—Traditional & Hazardous Pollutants*, both published by Ann Arbor Science Publishers.